D1429085

FINANCE AND TECHNOLOGICAL CHANGE

Finance and Technological Change

Theory and Evidence

Enrico Santarelli
Associate Professor in Applied Economics
Faculty of Statistics, Department of Economics
University of Bologna

First published in Great Britain 1995 by
MACMILLAN PRESS LTD
Houndmills, Basingstoke, Hampshire RG21 6XS
and London
Companies and representatives
throughout the world

A catalogue record for this book is available
from the British Library.

ISBN 0-333-63654-6

First published in the United States of America 1995 by
ST. MARTIN'S PRESS, INC.,
Scholarly and Reference Division,
175 Fifth Avenue,
New York, N.Y. 10010

ISBN 0-312-12852-5

Library of Congress Cataloging-in-Publication Data
Santarelli, Enrico.
Finance and technological change : theory and evidence / Enrico
Santarelli.
p. cm.
Includes bibliographical references and index.
ISBN 0-312-12852-5 (cloth)
1. Technological innovations—Finance. I. Title.
HC79.T4S2927   1995
338'.064—dc20   95-14913
                                                                CIP

HC
79
T4S26

10   9   8   7   6   5   4   3   2   1
04   03   02   01   00   99   98   97   96   95

Printed and bound in Great Britain by
Antony Rowe Ltd, Chippenham, Wiltshire

To my daughter, Lucrezia Maria

Contents

Acknowledgements

I have benefited from the advice of a number of scholars and friends who have read and discussed all or parts of the manuscript at various points in its development. In particular, I wish to thank Nathan Rosenberg and Nick van Tunzelmann for having initiated me into the fascinating field of the economics of innovation, Moses Abramovitz, Valeriano Balloni, Adrian Belton, Giorgio Gattei, Bronwyn H. Hall, Valery Katkalo, Alfred Kleinknecht, Enzo Pesciarelli, Paolo Pettenati, Roberta Piergiovanni, Marc Ryser, Alessandro Sterlacchini, Salvo Torrisi, Marco Vivarelli and Oliver E. Williamson for their many valuable suggestions. I also gratefully acknowledge the kind assistance received from the entire staffs of the Science Policy Research Unit at Sussex University, the Center for Research in Management at the University of California, Berkeley, the Department of Economics at Ancona University and the Department of Economics at Bologna University.

My appreciation goes to the Italian Ministry for University and Scientific and Technological Research and the National Research Council of Italy for their financial support during the period of time needed to complete this research.

Bologna   ENRICO SANTARELLI

Part I

Theory

1 The Role of Finance in Technological Change

1.1 FINANCE AND TECHNOLOGICAL CHANGE: THE NEGLECTED LINK

Theoretical analysis of technological change has traditionally focused on explanations of the interactions between production techniques, prices and quantities, and on interindustrial variations in capital intensity and market structure (see Gomulka, 1990). Less attention has been paid by economists to explaining the uneven patterns of technological change that characterise different industries, and to the feedback process set in motion by the interaction between technological and financial innovation. The dynamics of technological change, however, gained in interest during the mid-to-late 1970s and the 1980s, when technology drove a restructuring process that involved most industrial sectors in all the advanced countries. The mutual dependence of finance and technology is still given scant consideration by economic research, and there is therefore a need for more thorough analysis of the way in which financial resources are channelled towards innovative investments at the level of both the firm and the economy.[1]

The basic assumption of the present work is that technological change is *not only* a (technologically) self-generated process of scientific discovery, invention, innovation and diffusion. Historical evidence suggests that technological change derives, on the one hand, from the strategic and behavioural arrangements that characterise the interactions among firms, and, on the other, from the social and institutional arrangements that characterise the organisation and functioning of the entire economy at both the national and the international level. Accordingly, technological change can be described and explained by employing the conventional categories familiar to economists (firm level) and economic historians (socio-institutional level), with, at least for the economist, technical description of innovations limited to the strictly necessary. The economist and the technologist (engineer, physicist, etc.) may therefore work together more fruitfully than is usually the case, employing their respective tools of analysis and avoiding the sometimes counterproductive mutual encroachments that have typified innovation studies (in particular those by economists) in recent years.

3

At the strategic and behavioural levels, the size, the organisational form and the financial structure of innovating firms have changed radically over time. They also differ among industrial sectors, as well as among countries and regions. However economists have been principally concerned with the relationship between a firm's commitment to innovative activities (usually formal R&D) and its market power – one of the fundamental assumptions of the so-called 'Schumpeterian hypothesis' – while they have tended to neglect other aspects. In particular, little attention has been paid to how the nature of a firm's investment activities affects its choice among alternative sources of finance.

At the socio-institutional level, the way in which financial resources are gathered and employed to fund investment activities has also changed over time as a result of an innovation process similar to that which occurred in the field of production technologies. However, in spite of the importance of financial innovation within the overall process whereby new technologies are assimilated into the economic and social system – and in spite of J. A. Schumpeter's emphasis on its relationship with technological innovation – the recent literature on this subject has usually dealt with its effects on financial stability (see Minsky, 1982, and the Post-Keynesian approach), on the money supply (see Podolski, 1986) or on the effectiveness of monetary policy (see Kaufman, 1986). Only a few studies have systematically analysed the relationships between various developments in the finance process and the rate and direction of technological change.[2]

The aim of this book is therefore to bridge the gap between analyses of financial and technological determinants of long-run development on the one hand, and the studies of firm organisation and strategy on the other. It will single out and test empirically a number of hypotheses about the actual functioning of markets and the behaviour of firms when faced by a radical change in the technology being used in the production process.

From a macroeconomic perspective, the book contends that financial and technological innovations initially occur as autonomous events. Subsequently, financial innovations are adopted to finance the diffusion and widespread adoption of the new technologies because previous methods of finance are no longer adequate. As will be argued later, the patterns of economic growth and structural change are significantly affected by the interaction between technological and financial innovations.

From a microeconomic perspective, it is instead asserted that an innovating firm possesses better information than the *market* concerning the possible return stream generated by its specific investment activity. It is thus able to choose the finance structure that is better able to attract investors. As the following chapters will argue, the quality of innovative output at the firm level is conditioned by the firm's choice among alternative finance structures.

I take the perspective developed by evolutionary theorists with respect to the relationship, at the macroeconomic level, between new technology and the socio-institutional framework, whereas I rely on modern finance theory in analysis, at the microeconomic level, of the innovating firm's strategy and finance structure. Much progress has been made in both these fields of research in recent years. During the 1980s, innovation studies saw significant advances in explaining the formation of clusters of innovations, while, as regards finance studies, analysis of the role of informational asymmetries in determining the firm's financial structure made great strides forward.

In the case of innovation studies the revival of debate over technological change has created renewed interest in Schumpeter's analysis of the innovative process, with the emergence of the evolutionary (or neotechnological) approach.[3] In the case of finance studies, significant progress has been made since the early contributions of Kalecki (1937) and Modigliani and Miller (1958), which laid the basis for modern finance theory.

The evolutionary approach shares with Schumpeter the merit of having stressed the importance of innovative capability among the *determinants* of economic performance. In its turn, the modern approach to firm finance structure has improved understanding of the firm's choice among alternative sources of financing. Evolutionary theory is therefore helpful in analysing the long-run patterns of technological change and of its impact on the sectoral dynamics of economic development. Formal studies of firm financial structure instead aid understanding of how firms shape investment decisions to the aims of their investment projects. These studies demonstrate, in particular, that an innovative firm must choose the finance structure most likely to render successful its search for new innovations.

1.2 TECHNOLOGY AND INSTITUTIONS IN THE EVOLUTIONARY APPROACH

According to evolutionary theorists,[4] most of the R&D carried out within an industrial sector at any time is related to a well-defined behavioural routine that guides the evolution of the technology predominant in that industry during that given period. Accordingly, for each technology any advance seemingly follows a pattern that appears almost inevitable.

This assumption dates back to the definition of *technological convergencies* introduced by Rosenberg (1963, 1969), who pointed out how different industries converge to a common technological pattern and how *technological imperatives* guide the evolution of a given body of technological knowledge (see Santarelli, 1995).

Nelson and Winter (1982, pp. 258–9) extended Rosenberg's theory to assert that this 'frontier of achievable capabilities' defines a *technological regime* within which a series of *natural trajectories of technology* single out the direction of technical progress in specific fields and denote the results of scientific knowledge directly exploitable at the commercial level.

Dosi (1982, 1984) enriched this approach further, by introducing the analogous categories of *technological paradigm* and *technological trajectories*. In Dosi's words, a technological paradigm denotes 'a model and a pattern of solution of *selected* technological problems' (Dosi, 1984, p. 83) and embodies a series of technological trajectories that single out 'the pattern of "normal" problem-solving activity (i.e. of "progress") on the ground of a technological paradigm' (ibid.). Dosi (1982) – like Abernathy and Utterback (1978) in an earlier article – combines two important notions of the modern philosophy of science: that of *scientific paradigms* (introduced by Kuhn, 1963) and that of *scientific research programmes* (developed by Lakatos, 1970); in his view, the former is a synonym for technological paradigms, the latter for technological trajectories.

Sahal (1985) suggests instead the existence of *technological guideposts* and *innovation avenues*, and gives an account of innovation processes based on the systems theory. This view assumes innovation to be governed by a common system of evolution, such that:

> the process of technological development within any given field leads to the formation of a certain pattern of design. The pattern in turn guides the subsequent steps in the process of technological development. Thus innovations generally depend upon bit-by-bit modification of an essentially invariant pattern of design. This basic design is in the nature of a *technological guidepost* charting the course of innovative activity.
>
> There is an important corollary to the above proposition. It is that technical advances do not take place in an haphazard fashion. Rather, they are expected to occur in a systematic manner on what may be called *innovation avenues* that designate various distinct pathways of evolution. We may say that technological guideposts point to the innovation avenues just as the innovation avenues lead to technological guideposts. (Sahal, 1985, p. 71)

Technological trajectories (or natural trajectories of technology, or technological guideposts) are therefore the tangible results of the exploitation of basic technological knowledge in industrial activities, and as such they are more significant than technological paradigms (or technological regimes, or innovation avenues) for economic analysis of the production process in the strict sense. Within a technological trajectory, technological change is a continuous and cumulative process that follows a path-dependent, evolutionary pattern. Conversely the passage from one technological paradigm to another constitutes a dramatic change (a *discontinuity* in Rosenberg's terms) in the traditions of practice applied to the development of technological capability and denotes a major technological breakthrough.

But what is it that determines the bunching of innovations in technological paradigms and their diffusion? More importantly, what causes the upturn that brings about economic development?

1.2.1   From the Technological Regime to the Technological Style

Useful in answering the above questions are the definitions of the different bodies of technological knowledge prevailing in different historical periods and encapsulated by the ideas of *technology system*, developed by Freeman (1984), and of *technological style*, introduced by Perez (1983, 1985). These are particularly significant conceptualisations, as they explain how basic innovations initially, and the diffusion of secondary innovations that follows them, can generate long-run changes in economic activity.

According to Freeman, a deeper understanding of the birth, growth, and maturity of technologies would improve the analysis of the linkages between innovation and long-run economic growth and structural change. The first step is to demonstrate that the emergence of new technologies is not a smooth continuous process, but that it is extremely uneven over time. The second step is identification of what Freeman calls a *new technology system*, representing the 'cluster of innovations that is associated with a technological web, with the growth of new industries and services ..., and with new patterns of consumer behaviour'. New technology systems are particularly pervasive bodies of knowledge, and they determine a change in production techniques not only in their sectors of origin and main use, but also in most industrial and services sectors. Accordingly technological innovation drives long-run growth, through 'a rapid diffusion process which occurs when it becomes evident that the basic innovations can generate super-profits and may destroy older products and processes'

(Freeman, Clark and Soete, 1982, p. 67). Freeman completes his analysis of the role of new technology systems in long-run growth by asserting that no bunching of innovations takes place during economic depressions.[5] He shows that the swarms of innovations that really matter in terms of their expansionary effects are 'diffusion swarms', each of which represents a set of interrelated basic innovations – some of them technical, others social – concentrated very unevenly in specific sectors. There is thus a connection between the emergence of new sets of basic innovations and the onset of long and strong upswings.

Although Freeman's theory is a significant contribution to economic analysis of technical change, it fails to explain the upturn in long-run growth, and it does not clarify where clusters of innovations originate from and whether they are sufficient in themselves to generate an upturn (see Solomou, 1987; Tylecote, 1993). In this respect the most significant extension and integration of Freeman's contribution to the analysis of the role of technology in long-run economic dynamics has been proposed by Carlota Perez and subsequently refined by Andrew Tylecote (1993). These authors seek to complete Freeman's argument by explaining the causes of upturns in long-run economic growth and the extent to which technology is by itself able to generate an upturn.

Following Freeman, Dosi, and the French 'regulationist school' (see Aglietta, 1976; Boyer and Mistral, 1978), Perez attempts to fuse:

> the concepts of technological trajectories [Dosi] and technological systems [Freeman] and take them one step further. [She] suggest[s] that these notions are applicable to the analysis of the whole body of technology during relatively long periods, [and] propose[s] ... that behind the apparently infinite variety of technology of each long-wave upswing there is a distinct set of accepted 'common-sense' principles, which define a broad technological trajectory towards a general 'best practice' frontier. These principles are applied in the generation of innovations and in the organization of production in one firm after another, in one branch after another, within and across countries. As this process of propagation evolves, there is a prolonged period of economic growth, based on relatively high profits and increasing productivity. But, gradually, as the range of applications is more or less fully covered and when, through successive incremental improvements, the best practice frontier is actually approached, the forces underlying that wave of prosperity dwindle. As this occurs, limits to growth are encountered by more and more sectors of the economy, profits decrease, and productivity growth slows down. (Perez, 1985, p. 443)

The concept encompassing this process is that of *technological style*, a term that denotes a sort of paradigm for the most efficient organisation of production, and identifies the main directions in which productivity growth moves. A new technological style appears during a boom and causes rapid change in the organisation of production and in what Perez calls the *techno-economic subsystem* (or paradigm).[6] The development of a new techno-economic subsystem obliges designers, engineers, entrepreneurs and managers to acquire sets of rules and attitudes that differ substantially from those characteristic of the previous subsystem. However this technological revolution is not as smooth and problem-free as one might expect. In effect, a variety of problems may arise during a technological revolution because the modified techno-economic subsystem must coexist with a *social and institutional framework* that is better suited to the previous configuration. This 'mismatch' between the changing techno-economic subsystem and the unchanging socio-institutional framework is likely to lead to crisis and depression. This crisis entails a profound restructuring of the socio-institutional framework, and the introduction of social innovations and institutional arrangements that are more akin to the configuration assumed by the techno-economic subsystem after the emergence of the new technological style. An equilibrium configuration is reached only when and if the changed techno-economic subsystem 'matches' a new socio-institutional framework. It is this 're-match' that generates the upturn and allows faster economic expansion.

Tylecote (1993) extends these arguments even further. He cites in particular the various ways in which the mismatch between modified techno-economic subsystem and unmodified socio-institutional framework may provoke a crisis:

1. The old framework may block the diffusion of the new technological style and stop economic expansion, thus giving rise to a typical *depression crisis* responsible for severe social and political turbulence.
2. As the new technological style spreads in the economy and causes change in the techno-economic subsystem, the old socio-institutional framework may be sufficiently reformed so as not to block economic expansion. In this case, however, there is the possibility that a socio-political crisis will provoke wars or revolutions (*crisis of the upswing*).
3. The old socio-institutional framework may partially obstruct the diffusion of the new technological style. The consequence is a *mixed socio-political and economic crisis*, which does not determine disastrous events but only economic difficulties and socio-political tensions.

Tylecote emphasises that a new technological style (as defined by Perez) crystallises as the previous one becomes obsolescent and economic growth stagnates. Thus any new growth pattern needs, as its point of departure, a long and strong upswing in which an established technological style spreads rapidly and a new one crystallises at some point in the system. This upswing – as the new technological style profoundly alters the techno-economic subsystem – will lead in due course to a social and political *crisis of the upswing*, which will be relatively severe if the upswing is not preceded by a cathartic *depression crisis*, which provokes the social and political crisis enabling radical reforms to be introduced. According to the historical evidence, Tylecote points out that there have been no depression crises in the twentieth century. However the crisis typical of the twentieth century

> *being* a crisis of the upswing, ... [could] not be resolved by radical reform, ... [whose] outcome will be a socio-institutional framework which is little, if any, better matched with the new style than before. The stage is then set for a long and severe *downswing*, and for a renewed social and political crisis arising out of that downswing. This, going deeper, takes longer to resolve, but is in the end resolved with radical reform and a thorough 'rematch' of framework with style: the stage is now set for a long, strong upswing. In turn this boom brings about a crisis – less severe this time because the process of reform and rematch had been so thorough; as before, however, the crisis of the upswing leads to a modified, not fully-reformed framework, and the new style now crystallising interacts with this inadequate framework to produce another long, severe downswing and renewed social and political crisis. (Tylecote, 1993, p. 23)

Tylecote maintains that his 'differentiated Perez model' gives a satisfactory explanation for the emergence of mass production, or 'Fordist' technological style, which has characterised the twentieth century. The First World War and its aftermath coincided with a crisis of the upswing that was not resolved by radical reforms. Conversely, radical reforms were made after the long and severe world depression that followed the financial panic in 1929 and generated the 'longest and strongest international boom on record'.

Tylecote then summarises five main explanatory factors for the ups and downs in the world economy:

1. The succession of new technological styles, and the cycle of 'mismatch, crisis, and rematch' to which they tend to lead.

2. A population feedback, that is, the effects of economic growth and technological change on population growth and movement, and the effect of these demographic changes on economic growth.
3. A monetary feedback, that is, the effect of economic growth and technological change on real interest rates, and of real interest rates on economic growth.
4. An 'inequality' feedback, that is, the relationship between economic growth and technological change on the one hand, and between domestic and international inequality on the other.
5. The existence of a 'long cycle' in international relations, with the rise and decline of 'world powers' and 'global wars' ending with the appearance of a new world power.

Although this model provides a plausible explanation for the relationship between technological and socio-institutional factors in twentieth-century long-run economic dynamics, Tylecote acknowledges that it leaves a number of questions unanswered: (a) What causes the mismatch between a new technological style and an existing socio-institutional framework? (b) What are the effects of the different types of mismatch between technological style and socio-institutional framework? (c) How can a crisis be resolved? (d) How many reforms are likely to arise from a given type of crisis? As regards the socio-political mismatch between new style and old framework – which is the most important issue from the perspective adopted by this book – Tylecote asserts that 'inertia' and 'resistance' can explain a good deal:

> any major change in technology demands changes in the educational system, to produce workers with the necessary aptitudes and attitudes. It is scarcely likely that the reforms required will take place quickly enough to avoid some mismatch in this (economic) sphere [inertia]. ... Where entrenched *resistance* to change is most to be expected is in those aspects of the socio-institutional framework which are closely bound up with the vested interests and power struggles of social groups. (Tylecote, 1993, p. 71, emphasis added)

In Chapter 5 of the present book I shall attempt to explain how 'inertia' and 'resistance' have historically been responsible for the mismatch between financial institutions (and/or arrangements) – which represent an important component of the socio-institutional framework at any one time – and new technological styles, and how interactions among these forces have been most evident during periods of economic expansion. This, it is hoped, will provide at least a partial explanation of the relationship between technological and socio-institutional factors in long-run economic growth and structural change.

1.3   MODELS OF THE FIRM'S FINANCIAL STRUCTURE

At the firm level the innovative process is usually regarded as closely related to the R&D activities of new and/or established firms. According to modern finance theory, such firms require a certain finance structure if they are to fund, and implement, these R&D projects. One may therefore assume that all innovations that reach the stage of commercial exploitation result from long-term, 'D-oriented' R&D projects, and that these can only be undertaken by firms with the most appropriate finance structure (see Santarelli, 1991). Under these assumptions the innovating firm possesses better information than the *market* concerning the possible return stream generated by its R&D project. It is thus able to use its finance structure as a signalling device to attract investors (Riley, 1975; Leland and Pyle, 1977; Ross, 1977).

The proxy most commonly used when analysing a firm's finance structure is the *gearing ratio*, that is, the ratio between debt and the net value of the firm's assets. Finance theory comprises three main approaches to analysing the gearing ratio: the theory of increasing risk, the irrelevance theorem and the incentive-signalling issue. However it should be stressed that none of these approaches has analysed the relationship between a firm's financial structure and its innovative activities.

1.3.1   The Theory of Increasing Risk

In an article analysing the size of investment undertaken at a certain time by a given entrepreneur, Kalecki (1937) hypothesised that the cost of debt is a direct function of the gearing ratio. On the one hand, he suggested that the total risk is a direct function of the size of investment: in this case, the greater the investment the worse is the entrepreneur's bargaining position, since lenders perceive their position as endangered should the firm he/she runs perform badly. On the other hand the marginal risk also increases with the size of investment. As a consequence an entrepreneur who has taken on 'too much credit' faces interest rates that are higher than market ones. Thus interest rates represent fixed costs for the firm, irrespective of its bad or good performance.

Given the problems caused by a large debt, the amount of entrepreneurial capital is one determinant of a firm's ability to attract external funds. For example, if a firm issues new bonds whose value is too high compared with that of its own capital, the market may be reluctant to buy all these bonds, irrespective of their rate of interest. The obvious solution for the firm is to fix a level of investment that is lower than the amount of its own capital. In fact, even if only part of the firm's total capital is invested and the rest held in the form of securities, the entrepreneur will still obtain a net income from his/her capital. Conversely, if the firm's total capital is invested and things go wrong, the entrepreneur will be obliged to borrow at a rate of interest even higher than that usually charged on the market. This demonstrates that the risk of investment activities increases with the gearing ratio, and that the degree of entrepreneurial risk is an inverse function of the total amount of own capital that the entrepreneur is willing to invest. Moreover, since different entrepreneurs have different endowments of own capital, it is evident that the greater the entrepreneur's own capital the greater the amount of money he/she is able to borrow. It is consequently possible to observe firms of different size being started at any given time in the same industry. In fact the smaller the own capital of an entrepreneur, the smaller the 'normal' amount of credit that he/she can raise and the smaller the size of the firm that he/she can start.

This approach is valid if, when analysing a firm's finance structure, the differing types and purposes of its investment projects are not considered. In fact one may presume that differences in the future income streams of firms relate directly to the 'quality' of their investment outlays. One might for example assume that an innovative investment is likely to create a higher income stream than a routine one. In this case the potential outside investor perceives that the innovative firm – should its investment prove successful – will obtain a higher income stream and he/she thus relaxes his/her risk aversion (that is, his/her reluctance to buy the new bonds issued by the firm) irrespective of the firm's retained entrepreneurial capital. Kalecki's theory of increasing risk is particularly useful in explaining the function of venture capital[7] in the case of firms that have started their R&D activity by relying upon the entrepreneur's own funds and need further financing to complete the project. Under such circumstances the greater the amount of personal funds the entrepreneur–technologist has invested in the project, the easier it is to attract outside financiers for second stage financing.

1.3.2   The Modigliani–Miller Irrelevance Theorem

Modigliani and Miller (1958) developed an elegant model that demonstrates that a firm's value is not affected by its finance structure.

This implies that the choice of debt or equity financing does not influence the firm's choice among alternative investment projects. The conditions under which this theorem holds are: (i) a tax-free capital market; (ii) perfect certainty about future investment and the firm's return stream; (iii) the same degree of risk involved in alternative investment projects; and (iv) the irrelevance of firm size and history. The main arguments of this approach run as follows.

Consider an economy in which firms can finance their investments by issuing common stock or by resorting to bond issues as sources of funds. In equilibrium, the sum of the market value of the firm's common shares and of the market value of its debt are equal to the market value of all the firm's securities, that is, to the whole market value of the firm. Under this proposition the market value of a firm is independent of its finance structure.

This first proposition of the Modigliani and Miller theorem leads to a second one concerning the rate of return on common stock in the case of firms that have also incurred debt. In this case the rate of return on stock is a linear function of the gearing ratio, that is, it is 'equal to the appropriate capitalization rate for a pure equity stream ... plus a premium related to financial risk' (Modigliani and Miller, 1958, p. 271).

However, if one introduces the realistic hypotheses that information is asymmetrically distributed between the buyers and sellers of financial instruments (Akerlof, 1970), and that it is 'asymmetrically distributed also between those who make decisions (agents) and the theoretical beneficiaries of those decisions (principals)' (Greenwald and Stiglitz, 1992, p. 39), the Irrelevance Theorem does not hold. For example, if the discounted future income stream from a new investment is perceived to be higher by the management and lower by potential outside financiers, according to the irrelevance theorem the effect of undertaking the project will be the same, irrespectively of how it is financed. However, if the investment is financed out of retentions it will augment the value of current shares by the amount to which the discounted future income stream will exceed the cost of investment. Conversely, if it is financed by recourse to debt, the part of the future income stream that exceeds the cost of investment will be shared with outside investors (see Stiglitz, 1972; Fox, 1987). Thus managers interested in the long-term net cash flow of the firm will use retentions rather than debt to finance the firm's investment outlays.

1.3.3   Asymmetric Information and Firm Financial Structure

The literature on information asymmetries that ensued from Akerlof's (1970) paper on the 'market for lemons' considers the firm's finance structure to be a typical signalling device. The informational asymmetries considered in this literature relate to the inner features of the firm that is seeking financing. They have been treated by two main classes of models: those dealing with financial hierarchy (pecking order) and those dealing with signalling.

According to some authors (Donaldson, 1961; Myers, 1984, 1985) firms (a) prefer internal financing; (b) adapt their target dividend payout ratios to their investment requirements; and (c) if external financing is required they first issue debt (which is assumed to be the safest security), then hybrid securities such as convertible bonds, and finally equity. In this 'pecking order' (or financial hierarchy) theory informational asymmetries between investors and current firm insiders concerning the value of the firm's assets are such that equity is likely to be mispriced by the market. Thus if the firm plans to finance its projects by issuing new equity, the underpricing of this equity will be so severe that the new equity holders will capture more than the net present value of the project, thus creating a net loss for the incumbent shareholders. Underpricing can be avoided by financing the project with internal funds or non-risky debt, so that equity is the financial instrument of last resort (for an extension of this argument, see Harris and Raviv, 1990).

Myers and Majluf (1984) focus on a situation in which the firm does not have internal funds and is unable to issue any kind of risky or non-risky debt.[8] They develop a three-period model. At time t the firm possesses some fixed assets and is exploring an investment opportunity that entails further fixed costs in order to acquire some tangible assets. At this initial stage the value of the new assets is unknown to both the managers and the market. At time $t + 1$ the managers collect information on the value and features of the new assets. Since the market still lacks information concerning the investment opportunity, the problem faced by the managers at time $t + 2$ is whether to issue new equity or to forgo the investment. If they decide to finance the investment via new equity, the value (V^S) of the shares owned by incumbent shareholders becomes

$$V^S = \frac{P'}{P' + E}(E + a + b)$$

(1.1)

where P' is the post-issue price of the incumbent shareholders' shares, E is the amount of the newly issued equity (which is equal to the cost of the investment), a is the market value of the firm's total assets before the investment, and b is the present market value of the assets acquired to undertake the investment.

Conversely, if the managers decide to forgo the investment and no equity issue is made at time $t + 2$, the value (V^n) of the shares owned by incumbent shareholders is $V^n = a$. Thus the new issue and the investment are made if and only if $V^s \leq V^n$, that is, when

$$\frac{E}{P' + E} a \leq \frac{P'}{P' + E} (E + b) \tag{1.2}$$

Under the above conditions the incumbent shareholders acquire a part of the present value of the investment project, whereas the new shareholders capture a part of the present value of the assets installed before the new investment.

The 'pecking order' theory therefore relies on predictable events, namely the fact that internal funds are preferred to any kind of external funds, and by the fact that debt is always preferred to equity. Thus equity is a source of financing that is chosen only when other sources are not viable but the firm nevertheless decides to take up a promising investment opportunity.

The prototypes of the signalling models were those developed by Ross (1977) and Leland and Pyle (1977). These models differ in the signalling device adopted, which is debt in the former and the equity share that is retained by the manager in the latter.

Ross (1977) applied various of the arguments introduced by Spence (1973)[9] to the case of a risk-neutral manager with detailed knowledge of an investment opportunity who signals his/her perception of the quality of the investment through his/her choice of financing policy. Ross distinguishes between high-value (type A) and low-value (type B) firms – which respectively are representative of best performing and poorly performing firms – and assumes that high-value firms must choose a credible device in order to attract outside investors. High-value firms may simply claim to belong to type A, but this same announcement could also be made at zero cost by type B firms. Thus type A firms are forced to employ a more credible signalling device, represented by the firm's finance structure. According to Ross, the market perceives that a high gearing ratio is a cost to the firm, since it increases the risk of bankruptcy and reduces the manager's utility function. In addition, the value of the firm is perceived as a direct linear function of debt, whereas the cost of the gearing ratio is higher in type B than in type A firms, which can sustain higher levels of debt. By raising its debt level, the firm can obtain two opposite outcomes: on the one hand it increases the risk of going bankrupt; on the other it enhances its market value. Within a rational expectations framework, Ross shows that there is a separating equilibrium such that the firm's gearing ratio (and the risk of bankruptcy) increases along with the firm's quality, that is, that type A firms are more indebted than type B firms. In fact, in this case the manager in a type B firm faces positive costs if he/she follows the strategy of firm A's manager, since the increase in the present value obtained at time t (by cheating the market) is lower than the loss in the expected value at time $t + 1$ determined by the increased risk of going bankrupt.

In Leland and Pyle (1977) the manager is risk-averse and retains a share of the firm's equity in order to diversify his/her portfolio. For any increase in his/her equity share the manager incurs higher costs. However these costs are lower in the case of type A firms. Thus the gearing ratio of the firm and the share of the firm's equity retained by the manager are positively correlated with the quality of the firm. In this case the separating equilibrium is such that the manager in the type B firm has no incentive to choose the same strategy as the manager in the type A firm, since the cost of a concentrated portfolio is higher than the present value (as perceived in the market) of the firm.

In signalling models, the firm may therefore use its finance structure to induce a potential outside investor to take a share in its business, for example by resorting to external funds to finance a certain percentage of its investment outlays (Ross) or by buying back some of the equity it issues (Leland and Pyle). In fact entrepreneurs (managers) and outside financiers have different and sometimes divergent information on real investment carried out by firms, since the former have perfect knowledge of the quality of their investment projects, whereas the latter are only informed about the average quality of such projects. In the presence of significant informational asymmetries, the entrepreneur's (manager's) willingness to implement a certain project may be seen as a signalling device that stresses the quality of the project.

With respect to the irrelevance theorem approach, in the incentive-signalling model the assumption of perfect information and competition in financial markets is relaxed and the role of informational asymmetries between the entrepreneur (manager) and the outside investors becomes crucial. By assuming that entrepreneurs (managers) have information that potential outside investors do not, the choice of a certain finance structure may be considered as a signal to the market. From this perspective, the incentive-signalling approach enables a more thorough consideration of the *positive* function of the firm's finance structure as a determinant of investment activity. Accordingly, the analysis of the relation between firm finance structure and the undertaking of R&D activities in new technology-based firms and in established companies that will be conducted in Sections 4.3 and 4.4 will take up some of the suggestions of this approach.

1.4   TOWARDS A MIXED APPROACH

The chapters that follow adopt a theoretical and empirical approach that, despite their differences and respective limitations, combines components and methodologies taken from both the evolutionary approach to long-run economic growth and recent developments in the study of informational asymmetries between outside investors, who provide the necessary capital, and firm insiders, who control its use. Both approaches can be usefully employed to broaden the perspective originally developed by Schumpeter.

The evolutionary viewpoint enables one to conduct an in-depth analysis of the major influence exerted by new clusters of innovations on the behaviour of the entire economy, and to identify the combination of financial innovations associated with the emergence and diffusion of any new technological style. From this perspective, evolutionary theory is a useful tool of analysis for those who are seeking to explain the role of financial factors during the *diffusion* of new technologies.

Models of imperfect information in financial markets, in turn, are particularly useful when analysing the behaviour of firms adopting aggressive R&D strategies in order to produce new innovations. In this respect these models provide useful insights into the *invention* stage of technological change.

This book therefore uses the evolutionary approach mainly in its historical analysis – introduced theoretically in Chapter 3 and then conducted in Chapter 5 – of the relationship between technological and financial innovations during the long waves in economic development usually associated with the names of Kondratieff, Schumpeter and other authors. Comparison among the sources of finance employed during the expansionary phases of subsequent long waves reveals that certain financial instruments and institutions correspond to the diffusion of specific technological styles.

Moreover, two theoretical models of firm financial structure – dealing respectively with new technology-based firms and established corporations – are developed in Chapter 4 and used in Chapter 7 to analyse the financial strategies of innovative firms in the data-processing industry over the last two decades. These models and the subsequent empirical analysis show that, when undertaking aggressive R&D strategies aimed at developing a specific product innovation, entrepreneurial firms seem to rely most on financing contracts of the venture capital type, whereas under the same circumstances publicly held (quoted) and established companies prefer to issue new equity.

2 The Legacy of Schumpeter, and the Limits of the Schumpeterian Approach in the Analysis of the Financing of Innovation[1]

2.1 IDENTIFYING THE ISSUES

As stressed in the previous chapter, a theoretical analysis of the impact of financial institutions and the finance structure of the firm on the rate of innovation is still inadequate, and only limited treatment can be found in the writings of the fathers of economic science.

The role of finance in the overall process of technological change has in fact received scant consideration in both neoclassical and Keynesian theory. As regards neoclassical theory, this reductionist approach to the study of the financing of technological innovation is coherent with a theoretical tradition that neglects the impact of financial constraints on real resource allocation. In the Keynesian field, underestimation of the finance process can be instead traced back to Keynes' original formulation, which subsumed financial analysis under control of the money supply (for a critical assessment, see Greenwald and Stiglitz, 1992).

Although empirical work in the 1950s (Kuh and Meyer, 1959) suggested the importance of financial factors among the determinants of firms' investment behaviour, it was only in the 1970s that the study of the impact of financial structure on investment behaviour began to emerge with some salience in the theoretical literature.[2]

In the light of the above considerations, pre-Keynesian economic theory represents a surprisingly rich source of original insights, particularly the current of thought most outstandingly represented by J. A. Schumpeter[3] (see Dahmèn, 1984). This approach also comprises various developments in behavioural, institutional and evolutionary economics, most notably as a result of the work by Berle and Means (1932), Commons (1934), Simon (1955) and Nelson and Winter (1982) and their insistence on reconsideration of the role of entrepreneurial (and managerial) behaviour and the financing of entrepreneurial ventures.

20

Given that Schumpeter was the first of the fathers[4] of economics to stress that finance, together with entrepreneurial capabilities and technological change, represents one of the crucial factors in economic development (see Elliott, 1983), the present chapter outlines his approach to these matters. Nonetheless it should be noted that Schumpeter's contribution is used here only as a useful starting point for analysing the financing of innovation. In fact Schumpeter stressed the importance of financial factors in the financing of the *diffusion* of new goods and processes in economic activity, but he underestimated the problems connected with the analysis of the financing of *inventive activity* (science, industrial R&D and so on), which has become increasingly important during the present century since the introduction of R&D laboratories in large firms. Accordingly, in this chapter, after reconsideration of Schumpeter's original contribution to the analysis of technological change and the finance process (Section 2.2), an attempt is made to extend his argument to the case of the financing of inventive activity (Section 2.3).

2.2 FINANCE AND TECHNOLOGY IN SCHUMPETER'S EARLY WRITINGS

It is particularly useful to look at Schumpeter's early writings in order to gain a more thorough understanding of his analysis of the relationship between technological change and the finance process, and to single out its merits and shortcomings. These writings show quite clearly that Schumpeter's vision of the economic system was one in which static reequilibrating forces (which can be explained in terms of general equilibrium theory) are confronted by dynamic forces of disequilibrium represented by entrepreneurial capability and the availability of financial resources in the form of bank credit. By 'bank credit' Schumpeter means any funds raised by the firm in order to undertake those non-routine investments that, by fostering the process of innovation, provide the entrepreneur with a source of extra profits and monopolistic advantages.

Schumpeter sought to explain the forces of disequilibrium by constructing an original theory of development, whose general features are outlined in the writings of his 'German' period. This section therefore considers Schumpeter's first book, *Das Wesen und der Hauptinhalt der Theoretischen Nationalœkonomie* (1908) (henceforth *DW*), which has been largely neglected by economists, and the first edition of his *Theorie der Wirtschaftlichen Entwicklung* (1911) (henceforth *TED1911*).[5] As shown by Santarelli and Pesciarelli (1990), there are important differences between the first and the subsequent editions of *TED*, regarding both structure and contents.

The differences in structure lie mainly in Schumpeter's removal of the seventh chapter of *TED1911* and his rewriting of the second and the sixth chapters.[6] As regards differences of content, Schumpeter gives a new and original treatment to the concept of methodological individualism (Hodgson, 1986) and to his distinction between statics and dynamics; a treatment that provides a more complete understanding of the basis of his theory of economic development.

2.2.1 The Entrepreneur, the Banker, and the Financing of Innovation

In *DW*, Schumpeter gives a broad description of what he considers to be respectively the 'engine' of development and the 'ephor' of market economy: the entrepreneur and the 'banker'.[7]

These subjects are briefly mentioned in the part of the book devoted to bank credit. Here Schumpeter emphasises the 'effort' element and argues that it is by means of the creation of credit that 'the main stimulus for moving away from the condition of equilibrium and for making exceptional attempts is created' (*DW*, p. 420).

The figure of the entrepreneur is analysed in greater depth in other parts of *DW*, where Schumpeter lays particular emphasis on the roles of the 'dominant force', of personality, or of the 'energetic acts of will' (*DW*, pp. 351, 121–3, 567–8) in relation to the entrepreneurial function. Moreover Schumpeter constantly links these ideas to his belief that the activity of the entrepreneur is tied to entrepreneurial earnings and not to entrepreneurial wages, and that these gains cannot be explained within the domain of statics (*DW*, especially pp. 351, 420).

Thus, following the approach developed in *DW* and in Chapter 2 of *TED1911*, and having identified the main characteristics of so-called static subjects, Schumpeter adopts the concept of the 'energetic' type, whose chief property is 'energy of action'. A distinctive feature of this subject is that he introduces the 'new' into various activities by breaking with established routines – those usually adhered to by the so-called 'adaptive types' – and overcomes a wide range of social and cultural barriers.

In displaying his 'energy', the 'dynamic' type acts as the engine of history, the force of change.[8] Moreover, when discussing the way in which dominant personalities in various historical periods have imposed their will on the mass of 'hedonists', Schumpeter observes:

they can bend the majority to their purposes, even when the majority does not understand them or approve of them. They command, and the mass must obey. They impose new combinations, while the others keep to established custom. Such an organization, which imposes its will on the mass and forces it to act as it wants, existed throughout the Middle Ages and up to the dawn of the modern age. (*TED1911*, p. 185)

The power of this principle is such that it reverses the relationship between revolution and reaction: 'the progressive, or even revolutionary tendencies that might arise among the masses are of another sort: they are in fact a reaction against the actions of this minority. The mass is unwilling to introduce new combinations, and seeks to obstruct them or to ensure its greater share of the proceeds from them' (*TED1911*, p. 189).

Thus, in a general sense, Schumpeter describes two contrasting modes of behaviour embodied in two mutually contrasting subjects: the *energetic–dynamic* and the *hedonistic–static*.

The entrepreneur is the 'energetic' type transferred to the economic field. He is 'a leader of the kind skilled in economic affairs' (*TED1911*, p. 173). Schumpeter argues further that 'it is only in contemporary economy that the energetic type has developed to such a significant extent in the economic field as to constitute a special class of economic subject and be given his own name: entrepreneur' (*TED1911*, p. 171). However, and this marks a major difference between *TED1911* and the subsequent editions of the work, this is not to imply that the entrepreneur is 'simply a phenomenon of modern economy' (*TED1911*, p. 174), since his presence can be observed in previous epochs as well, although only in its 'essential features'.

According to *TED1911*, until the dawn of the modern age, entrepreneurs embodied two essential qualities: they were 'energetic types' and, like the feudal landlords, they exercised command power. These functions and powers were subsequently split by the division of functions, or if preferred, by the social division of labour. The arrival on the scene of the entrepreneur brought this process to an end, although in the meantime this same division of functions had reduced the power of command inherent in the entrepreneur's 'energetic' character. There therefore emerges an apparent contradiction in this particular aspect of Schumpeter's analysis: the impossibility of communication between the 'energetic' role of the entrepreneur and the static–hedonistic mass whose collaboration he needs. It is a contradiction, however, that enabled Schumpeter to draw up a theory of credit that provided a way out of the impasse in which theoretical economics found itself: the existence of a 'time lag between the deadlines that the entrepreneur has to meet in his tasks and his ability to fulfil them' (*TED1911*, p. 195).

It is in fact the banker who, by providing credit, furnishes the entrepreneur with the means necessary for the purchase of the factors of production. The banker's money restores to the entrepreneur that command power he has lost with the specialisation of his functions: '[the provider of capital] stands between the entrepreneurs and the providers of means of production. He is a phenomenon of development and appears only when the entrepreneur lacks power of command over the owners of labour and land. He enables the entrepreneur to become such, he gives him access to the means of production, he provides him with full powers for the fulfilment of his plans' (*TED1911*, p. 198).

This element acquires particular importance in Schumpeter's analysis, since it is through the figure of the banker that 'command power' is depersonalised, to become instead a good that can be purchased by all those endowed with the potential characteristics of the entrepreneur, irrespective of their membership of a certain class or social rank

> The first entrepreneur to break the hedonistic spell, which is to be found in every stationary economic system, has to overcome numerous difficulties. His action will meet with suspicion and with open or passive resistance. First the legal forms and technical conditions that he needs have to be created. *In particular, the financing of his enterprise is a completely new, unknown operation.* (*TED1911*, p. 431; emphasis added)

On the basis of this assumption, *bank credit represents a social legitimation of the introduction of new combinations* in a capitalistic economic organisation in which, unlike the feudal system, the 'energetic type' does not hold command powers over the factors of production. Thus the level of entrepreneurship within a capitalist society correlates positively with the level of financial resources that can be channelled towards innovative investments (the necessary although not sufficient condition for economic development), and it is this need by entrepreneurship for social legitimation that reveals Schumpeter's bias towards a macro-behavioural perspective (Etzioni, 1987).

Hence, *in modern economies it is only within a system with external financing that the whole of the existing entrepreneurial potential can be activated and innovative potential exploited to its full.* Otherwise there will be only limited numbers of entrepreneurs–innovators whose appearance will be due to quite random factors.

2.3 BEYOND SCHUMPETER: INVENTION AND THE FINANCE PROCESS

Schumpeter's contribution to the economic analysis of the relationship between entrepreneurs and bankers – as reconstructed in the previous section – is an important starting point for any attempt to develop a viable explanation of technological innovation as the result of economic investments allowed by the functioning of the socio-institutional framework. In effect, by stressing the role exerted by the finance process in determining the undertaking of innovative investments, Schumpeter fully recognises the crucial nature of the mechanism of credit creation emphasised earlier by Knut Wicksell. According to Schumpeter, innovative entrepreneurial activities require external financing both in 'heroic' and in 'trustified' capitalism: conversely, he regards inventive activity *strictu sensu* as an exogenous factor and ignores its financing.

In the light of the increasing importance of the structured inventive activity carried out in private and public R&D laboratories during the twentieth century (see Mowery, 1983), Schumpeter's assumption of the exogenous character of invention is a serious flaw in his analysis. His contribution is therefore a useful *starting point* for the study of technological change, but it only permits us to regard the transformation of an invention into an innovation as the work of entrepreneurs. Conversely, from a microeconomic perspective, the focus should also be on the features of *invention* activity and on the function of the firm finance structure during the undertaking of R&D and related investments.

Both in his interpretative analysis of economic development and in his theoretical premises Schumpeter displays a dichotomous vision. He sees development as taking place through a process of alternating 'technological impulses' – in explanation of which he elaborates his own original theory ('heterodox' Schumpeter) – and 'adjustment responses', which he explains by resorting to neoclassical theory ('orthodox' Schumpeter).

Schumpeter's 'heterodox' theory is characterized by:

1. His view of the introduction of major innovations as the basic impulse towards change and the cycle. Whereas changes in consumer tastes are regarded as being of minor importance, product and process innovations, the opening up of new markets, the finding of new sources of supply of materials, and the reorganisation of industry (the creation or dismantling of monopolies) are designated as major innovations. Thus the concept of innovation is given a wide range of

application that also includes organisations and the structure of markets.

2. Its identification of large-scale innovation as a new combination of factors introducible in indivisible form. And the twofold distinction between large-scale innovation as (a) *incremental innovations*, which accompany or follow the phase of adjustment ('swarming') and (b) *invention*, which usually precedes it, and *diffusion*, which follows it.

3. Its imputing to the ('energisch') entrepreneur–innovator of the capacity to introduce major innovations.

4. The ability of the entrepreneur to avoid budget constraints by resorting to credit and by using liquidity issued *ex novo* by the financial system.

5. The belief that the innovative process is discontinuous and that major innovations appear in 'swarms'.

6. The belief that in 'trustified' capitalism the innovative impetus of large firms will sooner or later begin to decline (Schumpeter, 1942).

This theory enables Schumpeter to describe the dynamics of the economic system and to contrast them with the endogenous tendency towards equilibrium that is typical of the state of circular flow, where traditional maximising behaviour adjusts the system towards the full utilisation of factors. Thus static processes are still interpreted in orthodox terms, dynamic processes in heterodox ones.

This interpretative scheme of Schumpeter's theory can be preserved as one moves from his early works on 'heroic' capitalism to his later writings on 'trustified' capitalism. In philological terms, it is possible to identify his celebrated article of 1928 as the connecting link between 'early' Schumpeter and 'later' Schumpeter.

As regards points 1, 2, and 5, in both the early and later Schumpeter, major innovations are still crucial to cyclical development – even if they are increasingly less the outcome of isolated 'acts of insight' and more and more the result of R&D in large enterprises or in small high-technology firms. Nevertheless, historically, cyclical fluctuations tend to diminish as mature capitalism loses its innovativeness and capacity for development. In a 'trustified' economy, according to point 6, there is a decreasing number of major innovations, and the changes brought about by technology tend to assume an incremental character.

Entrepreneurial functions, related to point 3, tend to dwindle, in the sense that anyone on the organisation chart of a large enterprise who shows innovative skill takes on the role of entrepreneur independently of his or her normal duties. Presumably this function has increasingly passed to R&D workers and product managers.

With regard to point 4, Schumpeter details the characteristics of the financial structure in the course of economic development. He identifies in 'heroic' capitalism a financial structure of pure credit where 'the essential function of credit consists in enabling the entrepreneur to withdraw the producers' goods which he needs from their previous employment, by exercising a demand for them, and thereby to force the economic system into new channels'. The financial structure is characterised by a central bank and by a banking system that, by creating credit, 'makes possible the carrying out of new combinations, [and] authorises people, in the name of society as it were to form them'.

Schumpeter distinguishes in 'trustified' capitalism a financial structure in which exist 'both the power to accumulate reserves and direct access to the money credit' (Schumpeter, 1928, p. 384). In this case the economy draws on a large number of different sources of finance (either internal or external) and not only on bank credit, that is, on external financing. Accordingly, despite these modifications in the financial structure, the introduction of innovations remains an entrepreneurial prerogative, and in 'trustified' capitalism, too, the entrepreneur is able to evade budget constraints.

However Schumpeter's model can be recast to highlight the crucial role of external financing in R&D and to introduce into economic analysis the stages of inventive activity leading up to the introduction of viable innovations. For analytical convenience we may assume that the only constraint on financial decisions by financial intermediaries is the profit calculus, and that as a consequence bank credit represents in equal measure credit created either by banks or by non-banking financial intermediaries. In these terms (Figure 2.1), entrepreneurial activities (EA) are the acts of skill and will ('animal spirits' in Keynes' term) that activate the vector of innovation (V of I) represented by the results of innovative investment (II) in 'D'-oriented R&D strategies. Within this framework the availability of bank credit (BC) – which can be taken to be representative of different types of external financing – is the necessary condition for the unfolding of the whole process and the implementation of the R&D project. The subsequent steps take the form of a new production pattern (NPP) characterised by new intersectoral linkages and relations, a new market structure (NMS), and the realisation of profits (or losses) represented by the total revenues minus the principal and interest that have to be paid to the outside financier. Conversely, when EA are primarily self-financed they normally lead to the undertaking of routine investments

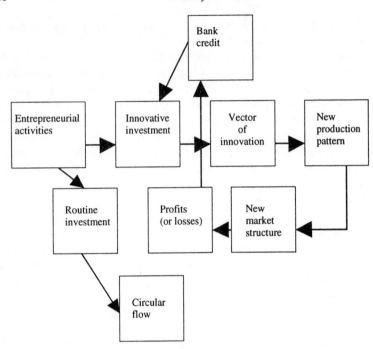

Figure 2.1   The financing of innovative activities in a post-Schumpeterian
perspective

(RI), which do not generate any virtuous cycle of innovation and do not modify the firm's and the economy's production patterns. RI are thus (according to Schumpeter) typical of the circular flow (CF), a static situation in which entrepreneurs make neither profits nor losses and in which their total revenues are employed entirely to acquire the factors of production.

With respect to *TED1911*, the crucial aspect is still the role of EA and BC, while the main difference lies in the endogenisation of invention as an economic variable relating to the actual financing of innovative EA.

Consequently a part of the uncertainty connected with inventive activity is shared between the entrepreneurs and the bankers. The former are the first decision makers, who submit to the latters' evaluation new research projects and new combinations of factors to be introduced in the productive process.

Thus, by modifying Schumpeter's arguments, it is possible to develop a much wider theory of the relationship between financial system, firm financial structure and innovation, one that enables a better understanding of the general process of technological change.

2.4 SUMMARY

This chapter has analysed the main aspects of Schumpeter's writings on entrepreneurial capabilities, technological change and the finance process. It has stressed that, although his analysis represents a useful starting point for study of the relationship between technological change and the finance process, it is largely unsatisfactory when the role of invention, as distinct from that of innovation, is taken into consideration. A reformulation of Schumpeter's theory has been proposed from this perspective.

Manifest in Schumpeter's early writings is his belief that static and dynamic forces determine the course of long-run economic growth and structural change. These forces can in turn be explained by general equilibrium theory – as is aptly pointed out in *DW* and *TED1911* – and by an original general theory of economic development that Schumpeter himself attempted to outline in *TED1911*. In his original theory, social and economic development depend on individual decisions and, in the case of modern market economies, on the existence of the social institutions such as the financial system that create the conditions for further development.

Once they have been modified to include *invention* among the most important economic variables, Schumpeter's suggestions can be employed both at the macro and the micro level. More aggregated analysis can be conducted to identify the long-run relationships between financial and technological variables, and their impact upon cyclical phases of economic development. Conversely, at the microeconomic level it is possible to discern the way in which a firm's financial structure adjusts to its technological activity; that is, how firms fund their innovative investments. These two distinct lines of analysis are pursued in most of the remaining chapters of this book.

3 Technology and Finance in Economic Growth and Structural Change

3.1 WHY FINANCIAL STRUCTURE MATTERS

On inspection of aggregated data and qualitative information, it appears that long-run rates of change in the financial and the technological structure follow a similar, although independent, pattern. In practice, financial innovations – which represent a part of the socio-institutional framework – occur as totally autonomous events and sooner or later are adopted in order to satisfy the financial requirements of new technological styles.[1] The discovery, development and commercial introduction of new technologies therefore appears to be closely related to the way in which innovative investment activities are financed. The process of technological change cannot be fully understood without identifying its relationship with the features of the financial instruments and institutions that characterise any given historical period. Here too, therefore, one may agree with Perez (1985, pp. 45–6) that the structural transformations that have historically helped to create the framework necessary for the emergence of a new mode of growth have generally also affected 'the organization of the banking and credit systems'.

This chapter is organised as follows. Section 3.2 analyses the most important institutional arrangements characterising the financial system in recent decades. Sections 3.3 and 3.4 introduce a partially new analytical approach to the study of the long-run relationships between technological and financial innovations. The basis of this approach is an extension of Perez's concept of 'style' to the field of financial institutions and arrangements.

3.2 MARKET-BASED VERSUS INSTITUTION-ORIENTED FINANCE PROCESS

Business finance can take the form of external or internal financing. External financing may consist of funds raised from capital markets, banks or other financing institutions, and from government, whereas internal financing includes retained profits and depreciation.

External financing may be either direct or indirect. It is direct when dealers in surplus funds acquire claims from debtors other than external financiers; conversely it is indirect when dealers in surplus funds acquire claims from external financiers (Santarelli, 1987a). Empirical analysis carried out by Gurley and Shaw (1957) demonstrates that from the end of the nineteenth century to at least the 1950s in the United States the amount of indirectly financed surplus units increased sharply, whereas the amount of directly financed surplus units diminished. These findings demonstrate that monetary expansion influences investment outlays, and they confirm the crucial role played by external financing in a fast-growing economy, where a process of financial deepening is a necessary condition for maintaining sustained rates of growth.

A high internal financing ratio may reflect a low level of capital formation or a misfunctioning of the financial system. For example the finding that industry in the United Kingdom is largely self-financed (some 70 per cent of funds are raised internally) is usually explained by the argument that low investment expenditures do not require major recourse to external financing. Nonetheless low capital formation may also stem from a shortage of long-term finance (as pointed out by Mayer, 1988, 1990). In fact a financial system able to provide only short-term finance generally fits well with an investment strategy focusing on routine rather than on extraordinary or innovative investments, the level of which by definition cannot exceed the rate of substitution of capital equipment.

However, as well as the problems of definition that remain, the distinction between external and internal financing is somewhat impressionistic. If one looks at the different ways by which industrial investment outlays are financed in the real economy, two different major systems can be discerned in the international arena in recent decades, each of them embodying some features of both internal and external financing (cf. Rybczynski, 1974): these are the 'market-based' and the 'institution-oriented' systems.

The former is based on the assumption that firms satisfy their financial needs by issuing equities and bonds or by resorting to retained profits. Within this 'market-based' system the stock market is the dominant financial institution, and the banking system serves corporate financing purposes only by providing short-term loans. This system is typical of the US and the UK and is generally found in the earlier developed countries (henceforth EDCs), where a large number of firms are traded on the stock market.

Conversely, in the presence of an 'institution-oriented' system, industrial growth is financed by long-term bank loans and under certain circumstances banks can own shares in industrial firms. In this prevalently credit-oriented system financial intermediaries and government are the main sources of investment capital for industrial firms, and the allocation of funds is rather discretionary because it depends on corporate past performance or on general political objectives. This system is typical of Japan and Germany among the EDCs, but it is mainly to be found in the recently developed countries (henceforth RDCs).[2]

Recent trends in international finance have altered the picture in many countries because the globalisation of financial markets gives depositors and borrowers access to global banking services on overseas markets. In particular, international financial integration and competition have necessitated the radical revision of national regulatory policies, which are by now unable to restrict the activity of national financial institutions without hampering their ability to compete in the international arena.

It is therefore not an easy task to distinguish between market-based and institution-oriented systems.

On the one hand, even in the EDCs increasing concern with the risk of stock market crashes (a risk that renders the market-based system more unstable than the institution-oriented system) has engendered a monetary policy that extends the control of central banks to non-banking financial intermediaries. In fact the recent wave of financial innovations has spread through the economic system because non-banking financial institutions of previously minor importance have assumed a much wider role. The traditional indicators of money supply are not significant when the function of money (or credit) creation is performed by non-banking financial institutions. It is thus very difficult to specify the current features of the finance process in many countries, and it is equally difficult to say whether it is more market-oriented or more institution-oriented.

On the other hand the willingness of national monetary authorities to sustain local financial institutions in international competition has brought about the deregulation of national financial systems, in the belief that the position of national banks in the international financial markets should be commensurable with the relative size of the national economy (Dollar and Frieden, 1989).

The result of these opposite tendencies is a continuous process of deregulation and reregulation, which increases uncertainty in local and international financial markets and gives industrial firms access to a much larger number of alternative sources of finance, either in the domestic or in the international market.

3.3 TECHNOLOGICAL AND FINANCIAL STYLES

The concept of 'style', which in Chapter 1 we saw employed by Perez with reference to those technological innovations that give rise to rapidly changing production functions for both old and new products, could also serve as an illuminating metaphor for financial innovations that cause change in the socio-institutional framework. With respect to technology in particular, the expression 'style' concisely denotes its nature as a cognitive structure devoted to solving significant economic problems.

We may adopt the above definition to suggest that the rapid diffusion of a new technological style – denoted by the formation of a diffusion cluster – usually takes place when general conditions in the socio-institutional framework are modified and when there is room for change at social, narrowly economic and technological levels. For analytical convenience, we may use Schumpeter's term 'upturn' for this situation, thereby distinguishing between *latent* and *active* technological styles. Although latent styles are embodied in currently available knowledge, they have not yet shown themselves to be of significant economic importance because they crystallise while active styles are becoming obsolescent. The usefulness of the knowledge base embodied in latent styles in solving specific economic problems is well known among technologists and scientists but not exploited at the industrial level because there is a mismatch between those styles and the old socio-institutional framework. For example, as we will see in Section 5.3.1, steam power was exploitable almost one century before the onset of the first industrial revolution, but it was only when this new source of power and the changing production requirements and demand conditions for the textiles and other industries came to coincide, along with radical reform of the socio-institutional framework, that a new techno-economic subsystem was able to emerge.

Active styles are instead sets of technological knowledge that underlie the expansionary phase of a new growth cycle. They spread through the entire economic system and determine the rate and direction of technological change in all industries. An active style emerges during the upswing of a new growth cycle and makes economic expansion possible.

In phases characterised by the obsolescence of active styles, when social conditions enable experimentation with new technologies, a number of innovations are expected to concentrate outside the obsolescent style(s). Indeed it is in this phase that potential basic innovations can be experimented with, and in some cases rejected. During expansions the selection mechanism identifies and selects the more promising among already existing basic innovations, and the active style is set in motion.

From an empirical viewpoint, we expect to observe a lower number of radical innovations within a single style whenever a latent style becomes an active one and the economy enters a period of expansion.

A third type of technological style may be presumed to exist. This can be called *dematured* style, one that is the outcome of the cumulativeness of styles: as new major technologies emerge, previous ones may reassert their importance. The meeting between an old, apparently exhausted style and an emerging one, for instance, can reactivate the former. This peculiar property of technological styles has consequences for economic analysis. In fact, during subsequent phases of capitalist development it becomes increasingly difficult to identify one leading technological style, since several connected styles develop, reach maturity and 'demature' simultaneously. In practice, a phenomenon of technological deepening takes place over time, as the techno-economic subsystem becomes much more complex and is characterised by several connected technological styles.

With regard to finance, the expression 'style' is particularly appropriate for financial instruments and institutions, since these seek to solve 'organisational' problems and define needs, although they do not necessarily always utilise scientific and technological knowledge for the purpose. Thus, just as it is possible to speak of technological styles as denoting new matrices of knowledge, so one may also identify financial styles that denote the characteristic features of financial instruments and institutions in the presence of well-defined interests, 'moral' attitudes, and overall economic pressures and tendencies.

The widespread diffusion of financial instruments and institutions is closely related to the course of technological change. On the one hand new technology – especially faster communications (telegraph, telephone and information technologies) – has sometimes made new financial practices and institutions possible; on the other hand most financial innovations are adopted to finance the emergence of new technological styles, which require a new investment pattern that only reorganisation of the overall financial system will permit. For instance, present financial institutions, as part of a crisis-ridden socio-institutional framework, cannot play any leading role in financing activities that belong to the new technological style, for at least three reasons:

1. Their conservativism and the custom of financing routine investment in traditional industrial branches.[3]
2. Their scant knowledge of the technical features and the feasibility of investment projects set up to explore technological trajectories belonging to the new technological style.

3. A geographical redistribution of investments brought about by the uneven diffusion of the new technological style among countries with different technological levels or devoid of an efficient financial system.

As a consequence, before the present financial institutions become aware of the feasibility of the new investment projects, and before they acquire the technical knowledge required to evaluate correctly the innovative investment programmes and develop the most suitable financial instruments, it is likely that new financial institutions will have been adopted. These latter will normally have formed autonomously as a *latent* financial style while the preceding financial styles predominated, but they assume an important function only when a new technological style emerges and when there is need for innovative, rather than routine, investments.

An environment of rapid technological change therefore gives rise to a radical reform of the financial system. But several other factors may be involved as well: for instance inflation, interest rates volatility, a shift in the flow of international saving and investment, growing competition in domestic and international financial markets, changing regulation in domestic financial markets, and so on. During a phase of structural and technological change, increasing demand for credit by 'developers of new technologies' may foster competition in financial markets, thereby reinforcing the ability of financial institutions to develop new financial instruments and expand their areas of business.

A major financial innovation is usually not created *ad hoc*, but stems from the full exploitation of an existing and previously underexploited financial instrument or institution. For example, as we will see in Section 5.3.2, the stock exchange market was not exactly a financial innovation when, during the boom of the 1850s, for the first time in modern history it became of particular importance to economic activity in most countries. In fact a stock exchange market had existed in the UK at least since the first half of the seventeenth century, although its role was marginal with respect to that of other financial institutions. As in the case of technological innovations, a time lag seems to exist between the invention and the subsequent exploitation and diffusion of a financial innovation.

From a microeconomic perspective, the main hypothesis to have emerged in the relevant literature is that the purpose of financial innovation is to slacken the increasing financial constraints that firms have to cope with. In neoclassical theory these constraints are essentially exogenous, although they are represented either by external or internal factors (Silber, 1975, 1983): (a) government regulation is a typical external constraint, ranging from increasing fiscal pressure to trade limitations, to complex labour and industrial relations regulations that obstruct the firms' use of traditional financial instruments and institutions; and (b) among internal constraints, close attention has been paid to those relating to firm size and market power.

As regards point (a) above, one may alternatively consider the cases of a price-setting, oligopolistic firm and that of a price-taking, small firm. In the former case the firm operates at its best practice frontier and sets both the price level and the yields volume. Its objective function being granted, this firm is expected to stimulate the creation of new sources of funds whenever environmental changes and new technological opportunities offer the occasion for the utilisation and development of new production processes and products that exploit a new technological style. Alternatively, the small, price-taking firm undertakes routine investment and borrows only the amount of capital it requires to achieve its production target. It may be presumed to borrow, among the funds supplied in the market by *new* financial institutions, only the amount that is strictly necessary to perform investments of an innovative and more profitable type. It will therefore continue to resort to its traditional lenders and to retentions for routine investments, and move to new financial institutions only if and when it decides to modify its routines and undertake aggressive, 'D'-oriented R&D strategies. Hence it follows that, from a microeconomic viewpoint, the stimulus to innovation is a response to increasing constraints: this response may be more or less costly, and in any case user attitudes to the cost of resorting to new financial institutions significantly affect the rate of diffusion of such institutions in the economic system.

From a macroeconomic perspective, financial innovation mainly takes the form of a financial response to 'credit crunches' (Sylla, 1982), in that the introduction of new monetary standards is usually the government's response to a crisis in the monetary system that stimulates a modification in the payment mechanism or in the monetary and financial instruments.

It is now possible to identify three main goals pursued by financial innovation in relation to technological change and the evolution of industrial structure: (a) to improve and adapt the lenders' ability to bear risk under the new conditions that are emerging in the techno-economic subsystem; (b) to improve – as does technological change – economic welfare, by strengthening the financial system's flexibility and its adaptability to corporate and individual needs, and by lowering transaction costs; and (c) to contribute at the margin to credit growth, thereby helping to determine the level of investments undertaken by industrial firms.

The onset of a new financial style thus permits greater bridging operations to support fixed capital, as well as changes in the organisation of the production system – such as the emergence of limited liability and the diffusion of giant firms – to spur technological innovation (for example by promoting investment in machinery that embodies technological change and through the R&D laboratories of large firms).

3.4 FINANCIAL MARKETS, TECHNOLOGICAL INNOVATION AND STRUCTURAL CHANGE

Economic analysis has investigated separately, at the macro level, the role of finance and technology in economic *growth*. On the finance side, empirical research carried out by Goldsmith (1955, 1969), monetary histories by Friedman and Schwartz (1963) and Kindleberger (1984), the theoretical framework developed by Gurley and Shaw (1960) and historical analyses by Cameron (1967) have provided points of departure for an impressive number of studies. Likewise the surge of studies on technological innovation during the 1980s and the 1990s showed that economic development can be promoted and reinforced by the availability of new technological alternatives (cf., among others, Freeman, Clark and Soete, 1982; Tylecote, 1993).

Compared with the number of studies dealing with either financial or technological innovations, relatively few attempts have been made to treat finance and technology as complementary determinants of economic *development* and structural change. The main contribution in this area – as stressed in Chapter 2 – is still Schumpeter's, although useful insights are also forthcoming from behavioural, evolutionary and organisational economics (cf. Commons, 1924, 1934; Williamson, 1988a).

To restate the crucial function of finance and technology in the analysis of economic development and structural change, one must first clarify the role of the financial system as a determinant of the production level and structure. This can be done by stressing two of its functions in particular.

Firstly, financial institutions are able to lift the restrictions on the accumulation process consequent on a divergence between income distribution and expenditure composition: they can transfer decision making concerning some of the resources owned by creditor sectors to debtor sectors. Thus savings can be allocated in order to stimulate investments in one sector rather than another, and, consequently the rate and direction of investment activities can be determined by the preferences of financial institutions.

Secondly, the financial system is able to influence the total money supply, according to the well-known principle of the credit multiplier. The obvious condition governing this process is that banking panic must not be triggered; that is, the situation must be arrived where economic agents do not withdraw their deposits at the same time and additional demand for credit arises from the economic system.

It is thus apparent that financial institutions can reallocate savings and create credit money as well, thereby indirectly controlling the total amount of funds devoted to innovative activity. In fact the financial institutions existing at a given time tend to finance a well-defined set of technological trajectories; these are the trajectories identified by the leading technological style, which is the one that diffuses greatest confidence among financial intermediaries. Thus we may assume that the degree of uncertainty associated with investments in latent technological styles stimulates the appearance of new financial institutions that are willing to face such an uncertainty.

In order to link the theoretical assumptions to the empirical investigation, we may assume that financial institutions and, in general, the techno-economic subsystem are driven by the profit motive. This amounts to saying that economic actors are persistently in search, within the realms of the feasible, for what may prove profitable, and that their search eventually leads to the emergence and subsequent development of new technological styles. In fact, during a depression both industrial firms and financial institutions lose profits, except for the most efficient of them, since these can survive the crisis by rationalising their internal organisation, or by exploiting the existing range of *technical* capabilities more efficiently.

Under these circumstances the perception that the situation is changing spreads through the socio-institutional framework and induces individuals and corporations to adopt new rules of business behaviour. This reform in the socio-institutional framework takes place because the new economic conditions require new routines whenever the adjustment process selects a few agents (individuals or firms) at the best practice frontier. As a result of imitation, the remaining agents shed their conservative attitude and are readier to undertake new and uncertain activities and productions. These new activities – which allow the actual assimilation of the new technological style in the techno-economic subsystem – are usually financed by new financial intermediaries and institutions that were of minor, if any, importance during the previous development cycle.

4 Asset Specificity, R&D Financing and the Signalling Properties of the Firm's Finance Structure[1]

4.1 FROM SCHUMPETER TO MODERN FINANCE THEORY

In Chapter 2 we saw how Schumpeter laid sound microfoundations for the analysis of the financing of innovative activities. Nonetheless Schumpeter's contribution is only a useful starting point for study of this phenomenon. As shown by the short survey conducted in Section 1.3, finance theory has progressed greatly in the last forty years. Accordingly it is much easier nowadays than it was at Schumpeter's time to analyse by means of elegant formal models the relationship between the types of investments undertaken by a firm and its finance structure.

Within the theoretical framework set out in Chapter 1, and following the intuitions of Schumpeter, this chapter therefore attempts to develop a theoretical framework that captures at the *microeconomic* level the relationship between the types of investments undertaken by a firm and its finance structure. It focuses in particular on the financing of specific 'D-oriented' R&D projects by an external financier of venture capital type and presents a theoretical description of the signalling properties of the innovating firm's finance structure. The theory developed in this chapter stands at the intersection of organisational economics, finance, and the economics of technological change. It draws on transaction costs economics (henceforth TCE), agency theory (henceforth AT), and the incentive-signalling approach (henceforth I-SA), lines of thought that offer many useful suggestions for the modelling of the relationship between finance structure, internal organisation, and investment decision at the firm level. In particular, both TCE and AT use an 'efficient contracting' framework, and both are interested in the study of managerial discretion (see Williamson, 1988b). Among the differences between the two theories, the most significant for the purposes of this chapter are that TCE takes the *transaction* as its elementary unit of analysis while AT uses the *individual agent*, and that TCE focuses on *ex post*, and AT on *ex ante* (with respect to the production process) contractual relations. Thus a combination of the two approaches allows study of the behaviour of *specific* agents when they are engaged in equally *specific* transactions. I-SA is, in turn, helpful in analysing the informational properties displayed by the firm's finance structure when firm insiders possess better knowledge than potential investors concerning high-quality investment opportunities (such as the R&D projects considered here).

TCE assumes that there is a sharp distinction between firms that commit themselves to specific-purpose, non-redeployable investments, such as the 'D'-oriented R&D considered here, and firms committed to general-purpose investments. Thus, since a specific R&D project involves asset specificity (as defined by Williamson, 1981), on the basis of Williamson's (1988b) arguments we may assume that in this case equity financing is the most efficient type of financing.

Likewise AT is employed for analysis of the behaviour of individual agents in a contractual relationship of the equity type, which is likely to arise in the financing of specific R&D activities when the agent has private information about the firm's R&D that the principal does not (see Myers and Majluf, 1984; Hart, 1988). The focus is particularly on start-up enterprises, for instance new technology-based firms (henceforth NTBFs), and previously established public firms (henceforth PEPFs).[2] In these cases informational asymmetries arise between those who control the R&D process and potential financiers (whether these are venture capitalists or shareholders), and AT is a powerful analytical tool when dealing with imperfect or incomplete information.

Moreover, the theoretical analysis set out in this chapter uses I-SA to capture the signalling potential of the finance structure of firms engaged in high-quality R&D projects that give rise to informational asymmetries between entrepreneurs in NTBFs or top managers in PEPFs and potential investors.

The above assumptions, drawn from TCE, AT, and I-SA, provide the basis for this chapter's principal arguments: (a) that the firm's finance structure is a crucial variable in the presence of *specific-purpose* investments entailing *asset specificity* such as the R&D expenditures considered here (Williamson, 1985, 1988a, 1988b); (b) that in this situation the firm's finance structure displays important signalling properties; and (c) that this firm enters into financing contracts of the equity type, which can be analysed as standard principal–agent problems.

The chapter is organised as follows. Section 4.2 uses the biotechnology and the data processing industries[3] to exemplify the theoretical arguments presented in the chapter. Section 4.3.1 reviews some features of the incomplete contract literature. Section 4.3.2 develops a standard principal–agent model of the relationship between a venture capitalist and an R&D-performing NTBF run by an entrepreneur wishing to undertake a specific-purpose investment. Section 4.4 outlines a model of the relationship between aggressive R&D strategies and the issuing of new equity in PEPFs. Section 4.5 makes some concluding remarks.

4.2   SOME STYLISED FACTS ABOUT R&D FINANCING IN HIGH-TECH INDUSTRIES

Contrary to conventional wisdom, most high-tech start-ups occurring in locational clusters of the Silicon Valley type are self-financed: entrepreneurs/technologists usually rely on money put up by themselves or by relatives and friends to implement the R&D projects that they confidently expect to succeed.

However, when the capacity to create internal funds is lacking,[4] it is more likely that finance of the venture capital type will be sought, rather than bank loans or any other kind of pure debt-financing (see Dean and Giglierano, 1990; Sahlman, 1990). Venture capital can be taken

> to be all risk bearing capital invested by a professional financial intermediary in promising companies or specific projects. Venture capital provides financial support, generally in the form of a participation in equity or an option to convert into equity. It is considered to have a strong risk-bearing character which focuses on high-tech and other technological industries with a high growth potential. In practice, however, venture capital firms are also interested in other promising activities. The relatively high risks are compensated for by the chance of high returns in the form of substantial capital gains. The venture capitalist not only provides capital for projects which he considers promising but also assists the management of the company, if necessary. The age of the company seeking the finance is not paramount although in practice younger companies tend to be major recipients. (EVCA, 1990, p. 9)

In accordance with the charter of the European Venture Capital Association, contracts of the venture capital type will be favoured under all circumstances in which the innovating firm *does not* possess much better information than the market concerning the possible return streams on its investment. The venture capitalist can in fact 'prove substantial activity in the management of equity or quasi-equity financing for the start-up and/or development of small and medium-sized unquoted enterprises that have significant growth potential in terms of products, technology, business concepts and services' (EVCA, 1990, p. 11).

Usually, young or newly established companies are the major recipients of venture capital, of which the most common type is *early stage financing*. However venture capital comprises several financing stages, each of which corresponds to a different stage in the development of a company. In particular, at least three more broad categories of venture capital besides early stage financing can be considered:

1. *Expansion*, that is, capital that may be used to finance increased production capacity.
2. *Replacement capital*, that is, the purchase of existing shares in a company by another venture capital firm or by a shareholder.
3. *Buy-out*, that is, capital that enables company insiders (or outsiders) to acquire an existing product line or the whole company.

In recent years, widespread recourse to venture capital has been an important feature of the early development and diffusion of information technologies and biotechnologies. As Section 7.2.1 will demonstrate, although invented many decades previously, venture capital as a financial innovation only became really important when employed to finance the new technological style of information technology and biotechnology.

The main advantage of contracts of the venture capital type is that they are set over a long period (at least five years), and that by signing such contracts entrepreneurs share the risk of the investment with outsiders. In fact the projects involved in transactions of the venture capital type are characterised by very high risk and a speculative nature. Thus, by looking for long-term commitment the entrepreneur/technologist signals to potential outside financiers the presence of asset specificity and non-redeployability in the project and his/her confidence that, when the R&D investment reaches maturity, the NTBF's return stream will be significantly high.

Conversely, in the case of pure debt contracts, even though they may be set over an equally long period, (a) stipulated interest payments must be made at regular intervals, (b) principal and interest must be paid at the loan expiration date, and (c) in the event of default the outside financier will realise differential recovery to the extent that the assets involved in the project are redeployable (see Williamson, 1991). The first point implies that the NTBF will be subject to liquidity constraints even before the project is fully implemented, and point (b) that failure to make payments on schedule will result in liquidation, even if the firm is economically viable.[5] As regards the third point, a deepening in the degree of asset specificity determines a decline in the value of pre-emptive claims, as a consequence of which the terms of debt financing will be adjusted adversely.

Venture capital has been widely resorted to in the US for development of NTBFs in circumstances where the entrepreneur's own funds cannot be relied on. Often the invention–development effort is undertaken by university researchers, who need financial backing to support their project (see Smilor, Gibson and Dietrich, 1990). For example Genentech Inc. of South San Francisco, one of the most successful US biotechnology companies, started up in this way when, in 1976, it began research on what would have eventually become its first commercial product: the human growth hormone produced by recombinant DNA technology.[6]

In other cases the innovating unit has been started up by entrepreneurs able to employ their own money for initial financing, but who need external funds for the further stages of the investment effort. For example Mitch Kapor used money he had raised from running other companies to start the Lotus Development Corporation (henceforth Lotus). When (in January 1982) he and his colleagues discovered that Kapor's personal funds were being depleted too rapidly, the decision was taken to seek outside funding in order to complete the design and development of new integrated software products.[7]

In the field of biotechnology it takes seven to ten years on average to develop new prescription drugs, at an approximate cost of about $200 million (at 1990 prices), while the innovation process takes place through exploration of new technological trajectories. The broad field of biotechnologies therefore represents a technological style in Perez's sense, whereas recombinant DNA technology is one of the most pervasive technological trajectories set in motion by this style. Genentech has undertaken R&D projects within the boundaries of this technological trajectory, alternatively focusing on product innovations such as synthetic growth hormones, tissue plasminogen activators (t-PA) and gamma interferons. The innovation process of this and similar firms therefore consists of the search for a specific new drug and entails a significant level of physical, human and even site asset specificity (Williamson, 1981). In effect each project employs technical and scientific instruments designed for a specific experiment, and human capital specially trained for that particular field of application.[8] Moreover such firms tend to cluster in a single excellence region as a consequence not only of historical accidents, but also of agglomeration economies and network externalities[9] (see Arthur, 1990).

In the field of software design the development of a new product that matches the needs of business and professional users, and is able to originate an integrated product line, requires the successful employment of new concepts, takes approximately two to three years and costs about $1.5 million (at 1990 prices). The first project of Lotus[10] was to create software that encompassed both spreadsheet and graphics capabilities, in this way enabling users to create and manipulate data and display the results graphically. The implementation of this project required significant human asset specificity and the combined efforts of Jonathan Sachs and Mitch Kapor, who had previously worked on the development of VisiPlot, one of the first microcomputer softwares allowing users to perform complex statistical analyses of data and to create graphs using the data. When the first round of venture capital financing began, Sachs and Kapor had already designed a series of increasingly powerful products, such as TRIO (which incorporated spreadsheet, graphics and word processing capabilities), SYMPHONY (which expanded the capabilities of TRIO to include database management and telecommunications features) and COMPOSER (a programming language system). Thus Lotus resorted to venture capital to continue its product development efforts and to relieve Kapor of the burden of financing the company out of his own money. The money raised by Sevin Rosen Partners (this being the name of the venture capital fund) was thus employed to accomplish the following ambitious goals:

- to become one of the top five independent suppliers of personal computer software within five years;
- to move into profit from 1983 onward;
- to devote significant resources to R&D;
- to achieve sales figures of $750 000 in 1982, $3 million in 1983 and $30 million in 1986.

All these goals were achieved. Indeed sales figures far exceeded the original target: total figures at Lotus amounted to $53 million early in 1983 and to $283 million in 1986.

Once the various stages of financing have been completed, the second step, either for start-ups or for previously formed companies resorting to venture capital, is to become publicly held companies. In effect, if the company does not go public or sell out to a large company, it is difficult for the venture capitalist to recover his/her investment. Thus the greater the amount of money the venture capitalist has invested, the more equity he/she will possess in the newly formed company. Genentech followed this procedure, and soon after the development of its first commercial product, it became a publicly held company. It is worth noting that the price per share, fixed at $0.72 when the first round of financing from venture capitalists began in April 1976, skyrocketed to $35.00 when shares were first offered to the public in October 1980 (see Sahlman, 1990). After Genentech went public in 1980, it achieved high rates of growth combined with high profitability. During transition from the NTBF to the PEPF form, almost all Genentech employees became shareholders, and stock options were introduced into the executives' compensation scheme.

A similar procedure has been followed by Lotus. In this case the first round of financing by venture capitalists began in April 1982, and after the initial public offering of October 1983, the price per share rose from $0.20 to $18. The very high rates of growth since achieved by Lotus (recently, in June 1995, taken over by IBM Co.) have stemmed mainly from the sale of its products incorporating spreadsheet and graphics capabilities, such as Symphony and 1–2–3. These have set the standards in a market later entered by other companies with products like Borland Quattro and Microsoft Excel.

To return to the case of Genentech, established companies displaying these features invest up to 40 per cent of their total revenues in new drug development. However the amount of money set aside for specific-purpose R&D investment is not always enough to cover all their expenses, since non-deferrable general purpose expenditures are also incurred. Thus the issue of new equity is a financing decision superior in efficiency to any kind of debt financing when the company undertakes an R&D investment entailing non-redeployability and specificity of assets. As Williamson (1988b, p. 580) states, 'confronted with the prospect that specialized investments will be financed on adverse terms, the firm might respond by sacrificing some of the specialized investment features in favour of greater redeployability. But this entails tradeoffs: production costs may increase or quality decrease as a result'. The alternative financial instrument available to firms seeking to avoid the tradeoffs entailed by debt financing is equity. This allows more managerial discretion than debt and is thus reserved for projects that are uncertain in nature and characterised by high asset specificity. In fact pure market exchange, to which debt is more akin, becomes more complicated when asset specificity is involved and some credible commitments are required to support the transaction (Williamson, 1975).

Genentech faced a similar situation in 1989, when it decided to resort to equity financing in order to raise the financial resources needed to implement its R&D projects. All the new equity was acquired by the Swiss pharmaceutical company Hoffman-La Roche, which thus holds 60 per cent of Genentech, which in turn has raised $500 million to accelerate the pace of its ongoing research activity and to undertake new special purpose R&D projects. The agreement with Hoffman-La Roche does not entail any shift in control: Genentech continued its activity as an autonomous publicly held company seeking to expand basic knowledge in the field of recombinant DNA technology and to develop new health-care products.

The agreement between Genentech and Hoffman-La Roche is of course a somewhat peculiar case in that the newly issued equity was subscribed to by only one major investor. Nonetheless it is representative of a general tendency in the biotechnology industry to fund special purpose R&D by issuing new equity. For example, Genentech itself has invested $15 million worth of equity in Glycomed Inc. for a research project to develop a carbohydrate-based drug. The rationale for this widespread recourse to equity financing in biotechnologies seems again to be risk-sharing combined with the consideration that, given the specificity of the assets involved in most R&D projects, a pure debt transaction would not be supported by credible commitments of any kind. Of course similar considerations also apply to the cross holdings and the equity links that characterise many cooperation agreements in this industry, which has shown a recent tendency towards bilateral relationships involving two or more firms committed to the development of a specific new product (see Orsenigo, 1989; Pisano and Teece, 1989; Teece, 1989).

4.3 A PRINCIPAL–AGENT MODEL OF THE FINANCE STRUCTURE OF A NEW TECHNOLOGY-BASED FIRM

4.3.1 Incomplete Contracts and Firm Finance Structure

The special case of R&D programmes considered in this section is that in which innovating NTBFs enter into long-term financing contracts of the venture capital type only in order to fund their R&D blueprints. In fact (a) they cannot make use of retentions because they are beginning their activity, and in any case the project requires an investment of funds in excess of their resources (this problem has been treated extensively by Stiglitz, 1974); (b) they cannot resort to short-term finance,[11] since this enables only routine investments, which reach maturity earlier than investments in 'D-oriented' R&D strategies; and (c) long-term debt is not worthwhile, because in the presence of asset specificity its terms are adjusted adversely for the NTBF.

NTBFs are influenced in the choice of an optimal finance structure by their requirement to enter into a financial contract that specifies the terms of future finance. The only contract fully displaying these features is an equity type one, which 'bears a residual-claimant status to the firm in both earnings and asset-liquidation respects [and] contracts for the duration of the life of the firm' (Williamson, 1988b, p. 580). Otherwise, if the innovative investment project has not reached maturity by the time a debt contract expires, problems will arise for both the firm and the outside investor. The former will be in an unfavourable bargaining position when renegotiating further loans, and the latter will run the risk of losing at least part of his or her funds should the firm fail to pay stipulated interest payments or go bankrupt (on the bankruptcy threat, see Aghion and Bolton, 1992. This may provoke conflict between the entrepreneur and the outside financier as circumstances change to the advantage or detriment of one or other party to the contract. A suboptimal solution to such conflict may be found by writing long-term contingent incomplete contracts (for a survey, see Hart and Holmström, 1987) of the venture capital type. These involve at least three different kinds of benefit.

Firstly, they provide a saving on transaction costs by establishing in advance what action is required of each party at different stages of the contractual relationship. The parties concerned therefore do not have to negotiate further short-term contracts and they thus avoid the problems arising from any change in the contractual conditions.

Secondly, long-term contracts enable the parties to avoid the difficulties created by further informational asymmetries arising during their relationship and which may affect their bargaining efficiency. This property is particularly significant in the situation considered in the present chapter, since it is one in which further informational asymmetries are likely to arise during the execution of the contract. In fact, by implementing the R&D project the entrepreneur–technologist improves his/her problem-solving capability and the competence gap with the outside financier consequently grows even wider.

Thirdly, a long-term contract also performs a screening function: it may attract the venture capitalist by offering high future equity interest if the R&D programme is successful. In this case the venture capitalist and the entrepreneur–technologist sign a contract prescribing future shareholding involvement by the former in all those circumstances that give rise to a tangible outcome from research activity. For example, if the venture capitalist puts up 70 per cent of the capital, he/she might require a 40 per cent equity interest.

4.3.2 The Model

According to the description of high-tech start-ups presented in section 4.2, and under the above considerations, let us consider an entrepreneur–technologist (or scientist) who, although he/she has access to a specific R&D programme, needs long-term finance in order to implement his/her investment project (Arrow, 1962; Mayer, 1988). This situation is likely to give rise to an R&D partnership where two parties – a limited partnership and a sponsoring company – enter into a contingent incomplete contract under which the sponsoring company performs R&D on a best effort basis[12] and the limited partnership provides the funds necessary for early stage financing.[13] The venture capitalist–limited partnership is assumed to be risk-neutral, and to be seeking all those investment opportunities that offer the highest profitability. For the sake of simplicity, it is also assumed that expected profitability is positively correlated with the degree of innovativeness.

The above is a typical case in which there are gains to specialisation; it is therefore one in which agency relationships between a risk-neutral venture capitalist and the innovating firm are likely to arise (Stoneman, 1987, Chapter 13; Holmström, 1989).[14] In fact the contract specifies in advance how the payoff is to be shared between the risk-neutral principal (venture capitalist) and the agent (entrepreneur–technologist); that is, it states what the equity interest for each party will be.

Let a denote a generic element among a given set of actions A (for example alternative ways to implement the R&D project) available to the agent. Let x then denote a verifiable outcome or a monetary payoff $x = x(a, \sigma \in \theta)$, resulting from the agent's choice of some action a and from the specific tangible and intangible assets σ involved in the project. In the case of a venture capital contract, outcome x is represented by the value of the NTBF's total equity when the project has reached maturity, the resulting new product has been introduced in the market, and the initial public offering is made. In practice σ is the idiosyncratic knowledge base upon which x depends, and it represents a portion of the technological trajectory θ, which contains a series of assets $\sigma_1 \ldots \sigma_n$ (with $\theta = [\sigma]$).

The problem now is determining how outcome x can be shared optimally between the agent and the principal in a contractual relationship where the principal puts up an amount of money y, enabling the agent to develop his or her R&D project from which a payoff x is expected.

Let y denote the amount of capital set aside by the principal and beneficial to the agent. The agent's utility function is $H(y, a)$ with $H_y \geq 0$, $H_a \leq 0$, and the principal's $U(x - y)$. The principal's utility function – which can be represented as a linear utility function – depends on the outcome of the inventive activity and the size of the payment y made to the agent to allow the undertaking of the R&D project, on $\sigma \in \theta$, and on a.

Venture capitalists – even if they are specialised in that given industry and have had personal training and/or experience in it – generally perform most of the analysis and investigation before investing their money. They gain access to the sponsoring company and make the investment only when they have obtained the information they want.[15] The empirical evidence on high-tech start-ups provided by MacMillan, Kulow and Khoylian (1990) indicates that venture capitalists are most involved in the financial and organisational aspects of the venture, at least compared with the entrepreneur–technologist. In particular, the lowest degree of involvement seems to be in those activities that concern ongoing research operations. In fact venture capitalists usually possess the basic technological knowledge required to ascertain whether exploration of a given technological trajectory will lead to the development of commercially successful products, but they lack the specific knowledge needed to implement a specific project. Their assessment of a specific R&D project depends on their perceptions of the normal rates of innovation in that industry, that is, of the leading technological trajectory.[16] Hence, in venture capital contracts of the kind considered here, the agent has greater competence in dealing with the tangible and intangible assets involved in the transaction than the financier.

Assuming that, as a consequence of this competence gap in problem-solving, the principal does not monitor the ongoing situation but only reviews the achievement of intermediate goals, we may demonstrate the existence of an optimal financing contract in which the sharing rules are functions of x alone. Following standard agency theory, the hypothesis of imperfect information concerns states of nature in which the agent's action a is chosen when the specific assets involved in the transaction are not known to the principal. In practice, imperfect information arises because the agent possesses a significant degree of knowledge concerning the non-redeployable assets $\sigma \in \theta$, whereas although the principal has sufficient knowledge of the technological trajectory θ, he/she has only limited knowledge (if any) of the specific assets that are involved in the transaction.

As in Holmström (1979), let $s(x)$ denote the agent's (NTBF's) share of x, and let $r(x) = x - s(x)$ denote the principal's (venture capitalist's) share. Pareto-optimal sharing rules to the agent, $s(x)$, are generated by solving the following programme:

$$\max_{s(x),\, a} \sum_{i=0}^{n} U[x_i - s(x_i)] \quad \text{where} \quad x = x(a, \sigma \in \theta) \tag{4.1}$$

which explains the variation of the discrete variable x. Accordingly, notation 4.1 can be substituted by notation 4.2, in which E denotes the expectational operator conditional on available information

$$\max_{s(x),\, a} \quad E\{U[x_i - s(x)]\}$$

(4.2)

subject to

$$E\{H[s(x), a]\} \geq \overline{H}$$

(4.3)

where $\overline{H}$ denotes the minimum level of the agent's expected utility, and

$$a \in \arg\max_{a^* \in A} E\{H[s(x), a^*]\}$$

(4.4)

where 'argmax' denotes the set of arguments that maximises the principal's objective function. Solution 4.1, subject to 4.3 and 4.4, is a second best solution, where notation 4.2 denotes that the agent is guaranteed the minimum expected utility, and the agent's objective function 4.4 takes account of the fact that the principal can only observe the outcome x but not the action a, since he or she does not monitor the agent's action. On the basis of the assumption that $x = x(a, \sigma \in \theta)$, the programme may be solved by taking the expectations in 4.2 and 4.3 relative to the distribution of $\sigma \in \theta$ (see Harris and Raviv, 1979).

In this model, the only persisting informational asymmetries – caused by a competence gap in problem solving – are those relating to the specific assets $\sigma \in \theta$ involved in the transaction. These allow adverse selection or, in terms of the TCE approach, opportunistic behaviour on the agent's side. To forestall the problems arising from adverse selection, the contract may be replaced by another one, independent of the agent's action a. The analytical procedure for dealing with this different contract, as described by Harris and Raviv (1979), overcomes the problems connected with opportunistic behaviour on the agent's side, even if the efficiency superior contracts that it entails allow fewer degrees of freedom.

Let us begin by considering two different contracts S_1 and S_2. The former is the standard contract as described above, where the sharing rules are a function of the agent's action a, a particular outcome x, and some specific assets $\sigma \in \theta$; the latter is a contract where the agent's action does not affect the sharing rules. Thus the problems of adverse selection are reduced, in the sense that they only depend on the informational asymmetries represented by the differing capability of the agent and the principal in dealing with the specific assets $\sigma \in \theta$. The signalling argument is more clearcut in contract S_2 than in contract S_1, since the outside investor has the guarantee that the NTBF will choose the action which, consistent with θ, enables achievement of outcome x.

PROPOSITION 1. Assuming $(S_1; x, a, \sigma \in \theta)$ to be the general form of the contract between the NTBF and the outside financier considered here, there is a contract $(S_2; x, \sigma \in \theta)$ such that any contract which depends on x, a, and $\sigma \in \theta$ can be dominated by a contract depending on x and $\sigma \in \theta$ only.

PROOF. As in Harris and Raviv (1979), let $a_1 = a(S_1)$, and $x \in \Omega$, with $\Omega^*(\sigma \in \theta) = \Omega(a_1, \sigma \in \theta)$, and then define the contract independent of action a as

$$S_2(x, \sigma \in \theta) = S_1[\Omega^*(\sigma \in \theta), a, \sigma \in \theta] - \Omega^*(\sigma \in \theta) + x \tag{4.5}$$

Hence, with V^A denoting the agent's choice of the action maximising his/her utility, and with $a_2 = a(S_2)$, we obtain

$$\begin{aligned} V^A(S_2, a_2) &\geq V^A(S_2, a_1) && \text{by definition of } a_2 \\ &= V^A(S_1, a_1) && \text{by construction of } S_2 \end{aligned}$$

In this case it is also true that

$$\begin{aligned} &\Omega(a_2, \sigma \in \theta) - S_2[\Omega(a_2, \sigma \in \theta), \sigma \in \theta] \\ &= \Omega(a_2, \sigma \in \theta) - S_1[\Omega^*(\sigma \in \theta), a_1, \sigma \in \theta] \\ &\quad + \Omega^*(\sigma \in \theta) - \Omega(a_2, \sigma \in \theta) \\ &= \Omega(a_1, \sigma \in \theta) - S_1[\Omega(a_1, \sigma \in \theta), a_1, \sigma \in \theta] \end{aligned}$$

Taking V^P to be the action maximizing the principal's utility, we obtain

$$V^P(S_2, a_2) = V^P(S_1, a_1) \tag{4.6}$$

In this case, since the realizations of $\sigma \in \theta$ and x are observable, a is inferrable *ex post*. The existence of such a contract has been proved by Harris and Raviv (1979).

If one wishes to solve all the problems arising from adverse selection, the importance of the specific assets $\sigma \in \theta$ in establishing the sharing rules should decrease even further, thus also reducing the degree of technological determinism implicit in the contract.[17] The problem arising over informational asymmetries relates to the role of $\sigma \in \theta$ in the establishing of sharing rules and compensation schemes for the agent. It can be solved in the manner set out by Harris and Raviv (1979), thus enabling consideration of a contract S_3 dominated by x alone, which is superior to S_1 and S_2 in terms of signalling properties. In fact by signing such a contract the NTBF (agent) guarantees that, although it possesses superior information, it will also act in the outside financier's (principal's) interest.

PROPOSITION 2. In the particular case where the NTBF is risk-neutral and the outside financier does not monitor the agent because of his/her limited knowledge of the technical features of the agent's action, any contract which depends on x, a, and $\sigma \in \theta$ will be dominated by a contract which depends only on x.

PROOF. As in the proof of proposition 1, let $(S_1; x, a, \sigma \in \theta)$ be a generic contract with $a_1 = a(S_1)$, $x \in \Omega$, and $\Omega^*(\sigma \in \theta) = \Omega(a_1, \sigma \in \theta)$, and define

$$c = E_{\sigma \in \theta}\{\Omega[a, \sigma \in \theta] - S_1[\Omega(a_1, \sigma \in \theta), a_1, \sigma \in \theta]\},$$

and let S_3 be a contract such that

$$S_3(x) = x - c \tag{4.7}$$

The principal is indifferent as to the choice between S_1 and S_3, because the risk-neutrality of the agent guarantees that he/she will put his/her maximum effort to the innovative activity. In this case, where both the principal and the agent are risk-neutral, the most efficient arrangement is one in which the principal offers the agent a contingent contract that gets the agent to internalise the effect of his/her effort decision (Kreps, 1990). The contract – which guarantees the agent higher rewards should he/she work hard – is such that the agent bears the full cost of displaying a lower level of effort. The agent will therefore compare among alternative R&D strategies according to his/her net expected utility. If the expected utility of the R&D activity that is associated with a higher level of effort exceeds that of the R&D activity requiring less effort, he/she is happy to work hard in any fashion. Hence the action maximising the principal's utility may be defined as

$$
\begin{aligned}
V^P(S_3, a_3) &= E_{\sigma \in \theta} U\{\Omega(a_3, \sigma \in \theta) - S_3[\Omega(a_3, \sigma \in \theta)]\} \\
&= E_{\sigma \in \theta} U(c) \qquad \text{with } S_3(x) = x - c \\
&= U[E_{\sigma \in \theta}\{\Omega[a_1, \sigma \in \theta] - S_1[\Omega(a_1, \sigma \in \theta), a_1, \sigma \in \theta]\}] \\
&= E_{\sigma \in \theta} U\{\Omega(a_1, \sigma \in \theta) - S_1[\Omega(a_1, \sigma \in \theta), a_1, \sigma \in \theta]\} \\
&= V^P(S_1, a_1).
\end{aligned}
$$

In his/her turn, the agent solves the following problem to maximise his/her utility:

$$
\begin{aligned}
V^A(S_3, a_3) &\geq V^A(S_3, a_1) \text{ by definition of } a_3 \\
&= E_{\sigma \in \theta} U^A[\Omega(a_1, \sigma \in \theta) - c, a_1] \\
&= U^A\{E_{\sigma \in \theta}[\Omega(a_1, \sigma \in \theta) - c], a_1\} \text{ by risk neutrality} \\
&= U^A\{E_{\sigma \in \theta} S_1[\Omega(a_1, \sigma \in \theta), a_1, \sigma \in \theta], a_1\} \\
&= V^A(S_1, a_1) \text{ by risk neutrality.}
\end{aligned}
$$

In this case the optimal contract, which is a second-best solution, is written when the sharing rules are such that the principal's share is independent of $\sigma \in \theta$, ignoring incentive problems. This equilibrium solution handles the problems connected with informational asymmetries between principal and agent, and the principal's lack of knowledge of the specific assets $\sigma \in \theta$. The existence of this contract has been proved by Harris and Raviv (1979).

4.4   A MODEL OF THE OPTIMAL FINANCE STRUCTURE IN THE CASE OF AN ESTABLISHED HIGH-TECHNOLOGY FIRM

As recalled in Section 2.3, it was Schumpeter himself who stressed that in the case of large firms, financing of the bank credit type is not the sole or main source of funds for innovating firms. Accordingly this section considers the case of large firms that issue new equity to fund R&D activities aimed at developing product innovations. On this basis and on the premises set out in Section 4.2, this section considers (a) the financial strategy of a large PEPF when it decides to implement an aggressive R&D project to develop a specific product innovation, and (b) the signalling mechanism set in motion by its top executives in order to raise the necessary funds. As was the case with NTBFs, a key element in the analysis is the private information about the R&D project possessed by the agents (in this case represented by the top executives of the firm).

In modern finance theory, the R&D costs faced by firms of the type considered in this section are those that result from planned search, the expected future return stream of which is discounted at a cost of capital that is conditional on the social risk inherent in that stream. Within this framework, individual portfolios or capital constraints do not present problems. However in the real economy, PEPFs annually allocate a limited amount of money to different projects, and the internal demand for funding is thus likely to exceed the amount set aside for R&D purposes in any particular year (see Holmström, 1989). Thus the issue of new equity gathers fresh funds for the company and allows the undertaking of long-term, special purpose R&D projects. This form of financing circumvents the problem connected to debt financing in the presence of asset specificity, and enables equity holders to act as risk sharers in the expectation of future rewards, represented either by high dividends or by an increase in the total value of the firm.[18]

Theoretical models in which capital structure is determined by agency costs belong to a tradition begun by Fama and Miller (1972) and developed in two classic papers by Myers (1977) and Jensen and Meckling (1976). According to Jensen and Meckling, when 100 per cent of equity is held by those not controlling the firm, its executives tend to slack. This tendency diminishes, however, when executives hold a share of total equity, because the greater their claim on the firm, the stronger their incentive to reduce slack. In this case, therefore, performance by top executives is closely related to their firm's performance, and stock options – which can be taken to represent the range of pecuniary and non-pecuniary benefits referred to by Jensen and Meckling – indirectly benefit the firm by improving its top executives' performance.[19]

In fact there is empirical evidence to show that the relationship in managerial firms between executive shareholdings and a firm's performance is usually positive and significant. For example Lewellen (1969) found that, in the early 1960s, the pay of top executives in large US corporations depended to a significant extent on the performance of their stockholdings. Lewellen showed that the ratio of the sum of after-tax stock-based remuneration, after-tax income and absolute after-tax capital gains to the after-tax fixed dollar remuneration of top executives rose sharply from 0.86 in 1940 to 6.9 in 1963. Similar results have been obtained by Lawriwsky (1984) in a study conducted on a sample of Australian companies.[20] More recently Gibbons and Murphy (1990) have shown that the revision of the pay schedule for chief executive officers, and the probability that they will keep their jobs the following year, are positively and significantly correlated with firm performance.

For analytical simplicity – since this section of the chapter focuses on the agency relationship between top executives and equity holders – we assume that top executives' decisions are fully accepted and implemented by R&D managers, that is, that no agency relationship exists between the central R&D function and the top executives.

Consider now a publicly held company belonging to a science-based industry with the objective function of maximising its total sales and profits. Assume that its top executives hold m per cent of the company's shares and are willing to undertake an aggressive R&D project (I_{RD}) in the belief that it may give the company competitive edge and increase its market share, thus enabling them to obtain additional stock options. The top executives can sell a fraction of the firm represented by newly issued equity to outside investors or to incumbent equity holders in order to collect the funds necessary to start the R&D project.[21]

This problem can be handled using a two-period model in which decisions taken in the first period affect the second-period return stream. Let us assume that the amount of the equity issued by the PEPF is the sum of the equity m issued at different times $m_1, \ldots m_n$ and that

$$\sum_{i=1}^{n} m_i = M_n$$

(4.8)

is the total amount of equity at time n. On starting its I_{RD}, at time $n + 1$ the PEPF issues new equities. The amount of total equity therefore increases from M_n to M_{n+1} and may be represented by the following sum

$$\sum_{i=1}^{n+1} m_i = M_{n+1}$$

(4.8)

Hence, if things go according to plan, the top executives receive (as in Ross, 1977) the following remuneration in period 2 (in terms of stock options F)

$$F = (1+r)\gamma_1 V_1 + \gamma_2 \begin{cases} V_2 & \text{if } V_2 \geq M_{n+1} \\ V_2 - L & \text{if } V_2 < M_{n+1} \end{cases}$$

(4.9)

where V_1 and V_2 denote, respectively, firm value in period 1 and period 2; L is the penalty to top executives if the firm goes bankrupt in period 2, after carrying out I_{RD}; r is the rate of interest; γ_1 and γ_2 are non-negative constants. The values of these constants differ from period 1 to period 2. In fact this model deals with logical time, and from this perspective the distance between period 1 and period 2 may be assumed to be long enough to determine a change in the constants.

Top executives maximise their objective function, as represented by the level of stock options F. Their main goal is therefore to implement I_{RD} and to maximise – by maximising the firm's total sales and profits[22] – the size of their stock options. Accordingly, notation 4.9 represents a signalling device, in the sense that M_{n+1} signals the firm type to outsiders and to incumbent equity holders. In practice the greater the difference between M_n and M_{n+1} in period 2, the clearer the perception by outside investors and equity holders that the firm is undertaking an innovative investment that will improve the firm's future return stream and value.

Let A^* and B^* represent two different PEPFs, with the former engaged in aggressive R&D programmes and the latter pursuing adaptive and imitative investment strategies, and let a and b denote the respective total return. Assuming M^* as the critical level of financing, taking F^* as the level of top executives' stock options when they own 100 per cent of the equity and F as the actual level, the result will be $b \leq F^* < a$ if $F > F^*$ – in which case the market will perceive the PEPF to be a type A^* firm – and $F \leq F^*$ if the market perceives the PEPF to be a type B^* firm. Corporation A^* will sell its newly issued equity at a face value such that $F^A \leq a$. Consequently firm value in period 1 will be

$$V_1 = V_1(F^A) = [a/(1+r)] \tag{4.10}$$

with r denoting the interest rate. Firm B^* will choose a face value such that $F^B \leq b$, and hence

$$V_1 = V_1(F^B) = [b/(1+r)] \tag{4.11}$$

At this point the stock options F of firm A^*'s top executives are

$$F^A(M^A_{n+1}) = \begin{cases} (\gamma_1 + \gamma_2) & \text{if } F^* < F^A \leq a \\ \text{and} \\ \gamma_1 b + \gamma_2 a & \text{if } F^A \leq F^* \end{cases} \tag{4.12}$$

while those of firm $B*$ are

$$F^B(M^B{}_{n+1}) = \begin{cases} \gamma_1 a + \gamma_2 (b-L) & \text{if } F^B > F* \\ \text{and} \\ \gamma_1 b + \gamma_2 b & \text{if } F^B \le b \le F* \end{cases}$$

$$(4.13)$$

The top executives may resort to a particular strategy in order to spread confidence in the market concerning expected returns on I_{RD}: they attract investors by increasing their total equity holdings, thereby signalling their confidence in the R&D project about to be undertaken.

Let m_{n+1} be the additional equity issued when the firm is planning to undertake an R&D project. Only an amount i of the new equity m_{n+1} such that $0 < i < m_{n+1}$ will be sold to incumbent and/or new shareholders, while an amount $(m_{n+1} - i)$ will be bought by its top executives. This is a typical reputation game (see Friedman, 1986), where a *closed loop* strategy is available to the participants. In fact the top executives decide to buy an amount $(m_{n+1} - i)$ of equity when they discover that potential outside investors are not interested in the new equity, whereas outside investors decide to buy the amount i of new equity on the basis of information that has become available after the beginning of the game, that is, when the top executives have bought back an amount $(m_{n+1} - i)$ of the new equity. The equilibrium solution referred to in the game is a Nash equilibrium.

PROPOSITION 3. Incumbent and potential equity holders lack confidence that the undertaking of an aggressive R&D strategy will be beneficial to the firm. In this case, top executives increase their equity share to spread confidence among equity holders over the firm's investment strategy.

PROOF. Let $V*$ be the face value of the firm's equity and

$$\sum_{i=1}^{n+1} m_i = (M_{n+1})V*$$

$$(4.14)$$

the total value of the firm's equity. If the issue of new equity is linked to an innovation-oriented R&D project, by buying a significant amount of new equity the equity-holding top executives signal to outsiders their confidence in the results of the new project. This will enable them to change their level of stock options in period 2, when extraordinary profits are achieved. The outsider and/or the incumbent shareholder will now be willing to pay the amount $(m_{n+1} - \lambda)V^*$ to buy the fraction of new equity $(m_{n+1} - \lambda)$.

In period 2, an increase in firm value takes place that brings direct benefit to the equity-holding top executives. This compensates them for the additional expenditures they have been forced to make in order to attract incumbent shareholders and/or outside investors to the new project.

4.5 SUMMARY

This chapter has employed standard static agency models to analyse the finance structure of firms engaged in aggressive, 'D-oriented' R&D strategies aimed at developing specific product innovations. R&D investment has been treated as a typical case of asset specificity à la Williamson, one able to affect the firm's choice of its finance structure. This has allowed the use of formal models to describe the financial strategies of both innovating small firms (NTBFs) and large corporations (PEPFs). These models correspond more closely to the main line of argument of the present work, namely (a) that the finance structure of the firm *does* matter in determining the choice among alternative investment (R&D) projects, and that (b) small firms and large firms resort to different sources of financing when undertaking aggressive R&D strategies. Moreover these models yield a more thorough understanding of the behaviour of Schumpeterian entrepreneurs and corporate top executives, and they also enable the analyst to employ qualitative data from the firm when investigating the set of processes and decisions involved in the undertaking of R&D projects.

Regarding the financing of aggressive R&D carried out by NTBFs, the chapter's major conclusions relate to two kinds of optimal contracts: those that allow R&D financing contracts to be developed in which the sharing rules are independent of the agent's action; and those in which the sharing rules are independent of both the agent's action and the specific assets involved in the transaction. Regarding PEPFs undertaking aggressive R&D, the chapter has shown that in this case too equity represents the optimal financing strategy, and that top executives increase their equity share as a signalling device to attract new and incumbent shareholders. These theoretical findings will be tested empirically in Chapter 7 against qualitative information and statistical data regarding, respectively, the financial strategies of NTBFs and PEPFs belonging to the data processing industry.

Part II

Evidence

5 The Long-Term Dynamics of Financial and Technological Styles: How Close a Relationship?

5.1 INTRODUCTION

This chapter analyses the long-term dynamics of financial and technological styles from both a 'quantitative' and a purely 'qualitative' perspective. Section 5.2 deals with quantitative aspects. In particular, Section 5.2.1 surveys some empirical implications of the Kondratieff long-wave hypothesis and points out the main methodological problems encountered by previous researchers. Section 5.2.2 addresses the long-term dynamics of, and the relationship between, two time series regarding *basic* technological innovations and certain important financial innovations between 1691 and 1971. Analysis is conducted of the growth rates, the mean value and the variance in the distribution of both types of innovations during each upswing and downswing phase of different chronologies of Kondratieff long waves.

From a qualitative point of view, Section 5.3 then examines the relationships between five groups of technological and financial styles characteristic of five alleged Kondratieff long waves since the 1790s. In this connection, Section 5.3.1 treats the interaction between the exploitation of water power and *stationary* steam power on the technological side and the use of merchant capital in industrial activities on the financial one. Section 5.3.2 analyses the relationship between the railways' technological style and the stock exchange financial style. Section 5.3.3 considers the relationship between the chemicals and other technological styles and the adoption of the international gold standard. Section 5.3.4 looks at the relationship between the 'Fordist' technological style and the financial style based on the crucial function of central banks in guaranteeing monetary and financial stability. Section 5.3.5 analyses the role of non-banking financial institutions, such as venture capital firms and pension funds in the early diffusion of information technology and the biotechnology industries. Finally, Section 5.4 summarises the main findings of the chapter.

5.2   PATTERNS OF TECHNOLOGICAL AND FINANCIAL INNOVATIONS

5.2.1   The Long Waves Hypothesis Revisited

In Section 1.2.1 it was asserted – in accordance with Perez (1985) and Tylecote (1993) – that a new technological style is likely to spread in the economy when the old socio-institutional framework has been sufficiently reformed and therefore no longer obstructs the assimilation of new technology into the techno-economic subsystem. Hence it follows that financial instruments, institutions and institutional arrangements – as part of the overall socio-institutional framework – must be *sufficiently reformed* or *drastically changed* if they are to permit the faster *diffusion* of a new technological style already under way and the exploration of the technological trajectories that it sets in motion. Adopting Perez's (1985) original perspective, it might therefore be argued that

> the transition to a new techno-economic regime [subsystem] cannot proceed smoothly,... mainly because the prevailing pattern of social behaviour and the existing institutional structure were shaped around the requirements and possibilities created by the previous paradigm.... Under these conditions, *long wave recessions and depressions* can be seen as the syndrome of a serious 'mismatch' in the socio-institutional framework and the new dynamics in the techno-economic sphere. The crisis is the emergency signal calling for a redefinition of the general model of growth (Perez, 1985, p. 445, emphasis added).

According to this view, the fluctuations observed in economic growth and the overall process of structural change reflect fluctuations in both technology and the prevailing socio-institutional arrangements.

In fact, after a wave of optimism during the 1950s and the 1960s and the conviction that the industrialised countries had entered a golden age characterised by stable growth, the crisis of the 1970s redirected attention to recurrent cyclical economic phenomena. This resurgence of interest in the instability of the growth path in turn focused renewed attention on the Kondratieff (1935) long-wave hypothesis, which posits the existence of regular economic fluctuations with a wave length of 45–60 years. The phenomenon of long waves described by Kondratieff and soon accepted by Schumpeter (1939) then lapsed into oblivion. However in recent years it has been reexamined by various authors in an attempt to identify the determinants of the long-term dynamics of technological development.[1]

Kondratieff described two different kinds of long wave:

1. An irregular sinusoidal wave with a length of approximately 50 to 60
    years – divided into two phases of upswing and downswing lasting
    about 25 years each – which could be identified in price variations
    prior to 1920.
2. A trend variation wave lasting an average of 50 to 60 years with a
    high-growth path significantly higher than the low-growth path.

Although he had little to say about the determinants of long waves,
Kondratieff identified various descriptive characteristics of the
phenomenon, five of which deserve particular mention:

1. Long waves are international.
2. In all examined trends in prices and production, long waves change
    direction at the same moment.
3. Years of prosperity predominate during upswings, whereas years of
    depression are more numerous during downswings.
4. Innovations (although Kondratieff called them 'inventions') tend to
    cluster during downswings and become widely adopted during the
    next upswing.
5. Wars and revolutions are more likely to occur during upswings than
    during downswings.

Kondratieff, however, was not the originator of long-wave theory, since it
had in fact been previously developed by a number of Dutch economists
between 1915 and 1932 (for a review, see Kleinknecht, 1987, ch. I).
Among them, van Gelderen – writing under the pseudonym J. Fedder
(1913) – was the first to draw 'attention to the tendency for long-term
price movements, interest rates and trade fluctuations to follow a cyclical
movement lasting about half a century' (Freeman, Clark and Soete, 1982,
p. 19). Van Gelderen extended his long-wave theory even further, to
include the hypotheses that (a) each upswing in a long wave is driven by
the rapid expansion of one or several rapidly growing sectors and
ultimately made possible by the availability of cheap loan capital, and
(b) *the long wave is accompanied by credit expansion and increasing
tensions in capital markets*. Van Gelderen's idea of the cyclical patterns of
technological innovations and of their ability to generate new industries by
means of an allowing mechanism represented by the financial system was
an evident stylisation of the idea originally advanced by Schumpeter and
sustained by this book: namely that the interaction between technological

and financial factors is one of the determinants of economic growth and long-term fluctuations.

Van Gelderen, Kondratieff and Schumpeter are not the sole proponents of the long-wave hypothesis: van Duijn (1983, p. 163) and Bieshaar and Kleinknecht (1983) list a total of twelve different chronologies of long waves, each of which can be associated with a different author(s)[2] (see Table 5.1). Bieshaar and Kleinknecht (1983) and Kleinknecht (1987) have tested some of these chronologies, identifying that by Mandel (1980) as the most suitable for studying the phenomenon of long waves on a world market scale, and as the least restrictive regarding the regularity of long waves.

Despite the various efforts of recent years to test the theory of long waves empirically, at least two major methodological problems still remain to impede thorough verification of the existence of such waves:

1. The time series available for most economic phenomena are too short for reliable statistical analysis.
2. Economic time series in general do not meet the requirements of stationarity; they must therefore be detrended using the most neutral statistical method.

As regards the first problem, rather unsatisfactory time series are available on innovative output, mainly derived from patent statistics, but almost no data at all on financial innovations, since a patent system for these events does not exist.

With respect to the second point, students of long waves have adopted various methods to detrend the time series employed in their analysis. For example Kuczynski (1978) used spectral analysis, but this method does not clarify whether and to what extent the results obtained depend on the trend model chosen by the author. From a different methodological perspective, Mandel (1980) and van Duijn (1979) computed and compared average growth rates for various subperiods in long waves, but they failed to test for the significance of observed differences in growth rates in the relevant phases. More methodologically sound analyses have been carried out by Solomou (1987) and Bieshaar and Kleinknecht (1983).

Solomou (1987) develops a phasing of trend variations based on the Juglar cycle as the unit of trend phases. By using estimates of gross domestic product, indices of industrial production and productivity data, he first computes the mean geometric growth rates for the period 1856–1973. This analysis shows that the relevant variables have not been steady over long periods throughout history. After testing variations in the growth paths by introducing a dummy variable into the relevant autoregressive scheme for the growth variable, Solomou concludes that

> the evidence rejects the Kondratieff wave phasing of post-1850 economies.... In fact, non-steady growth over long periods has characterised the world economy throughout the period 1856–1973; Nevertheless, the observed variations do not follow a Kondratieff wave pattern.... the structure of growth has not remained constant over time. Britain and Germany traversed from a Kuznets swing mode of growth variations before 1913 to a 'trend-acceleration' pattern of growth in the twentieth century. France followed suit after 1950. Even in America, where a long-swing pattern of growth was observed for a much longer period, the structure of these swings variations changed significantly. The amplitude of the 1930s downswing was much higher than any other period. The idea that economies grow along a regular cycle is not valid either for the Kuznets swing or longer-run growth variations. In much of the literature episodic changes have been mistaken for cycles of economic growth. (Solomou, 1987, pp. 61–2)

Bieshaar and Kleinknecht (1983) conceive long waves as a succession of longer periods of accelerated (upswings, or 'A-periods') and decelerated (downswings, or 'B-periods') growth, and assume that, if the long-wave hypothesis is relevant, the alleged upswings should display average growth rates of GNP or manufacturing production that are significantly higher than those of the preceding and following downswing, and vice versa. These average growth rates are computed for time series of gross national product and manufacturing production from some different countries, and for two series of world industrial production.[3] From a methodological viewpoint, Bieshaar and Kleinknecht assume that the transition from A periods to B periods and vice versa is not subject to erratic jumps in the absolute levels of the variables, and they impose the restriction on the trend estimates that in the transition years (that is, the 'peak' and 'trough' years of the long waves) the estimated values of the trends for the preceding and the following periods must equal each other. The results of their analysis differ from Solomou's inasmuch as they partly support and partly contradict the long-wave hypothesis. According to Kleinknecht:

> On the one hand, important series such as those on world production, or the data for Great Britain, France and Germany give no support for long-term fluctuations of the Kondratieff type during the period before

1890; on the other hand, the Belgian data show a highly significant long wave pattern from the 1830s onwards.... Furthermore, evidence of long waves during the period before 1890 comes from the Italian and Swedish data, although for shorter periods. (Kleinknecht, 1987, pp. 31–2)

Nevertheless, both Solomou's and Kleinknecht's analysis deal with indicators of general economic activity that are better suited than innovation output indicators to treatment by detrending statistical methods. In particular, the methodologies introduced by Solomou and Kleinknecht cannot be straightforwardly applied to analysis of 'qualitative' data on technological and financial innovations.

5.2.2 Nature of the Data and Empirical Results

With respect to technology, patent statistics cannot be taken as reliable indicators of the output from innovative activities. Although widely employed in empirical studies, they raise at least five main problems (see Santarelli and Piergiovanni, 1994): (1) the economic value and the technological level of each patent are highly heterogeneous; (2) the preference for bundling claims together in one or more patents varies widely from one country to another; (3) not all innovations are patented; (4) not all patents become commercial innovations; and (5) firms of different size have different propensities to patent.

Their limited heuristic value notwithstanding, patents represent the most viable source of historical innovation data. In this respect, the largest source is Baker's (1976) list of 363 *significant* patents since 1691 onwards.[4] This source had its origins in a small collection of references assembled by the staff of the British patents enquiry desk at the Science Reference Library. It was subsequently augmented when a search conducted on a large number of textbooks, encyclopaedias and journals permitted identification of the patents relating to significant innovations.[5] Compared with other patent-based counting procedures, Baker's has the merit of using a reliable method for the ex-post evaluation of the economic value of each patent included in the data base. In effect, this indicator is constructed by a process that selects only those successful patents that have brought significant progress in their field of application and given rise to 'swarming'. However this method also suffers from serious limitations. In particular, it encounters the problem listed under heading (4) above, since it reports only the year in which the invention was published in Patent Office leaflets,[6] but not the date of the market introduction of its corresponding patent. Furthermore Baker provides no information on significant innovations that, although placed on the market, have never been patented.[7]

Since Baker's data only relate to original inventions, they cannot be used to analyse the diffusion process that – after a gestation period that varies greatly depending on purely technological, entrepreneurial, demand, and environmental factors – follows the introduction of *basic innovations*; a diffusion process characterised by a swarm of *incremental innovations* or *minor improvements*. Thus, having established that the emergence and assimilation of new technologies is not a smooth process, it is very difficult to assert that (a) as hypothesised by both Kondratieff and Schumpeter, basic ('heroic') innovations tend to cluster during depressions rather than booms, and (b) the lead times from invention to innovation are shorter during periods of depression or periods of boom.

As regards point (a), it is highly likely that basic innovations (although it would be more correct to call them 'inventions') are also developed in periods of stagnation and depression. Therefore they cannot be considered a phenomenon typical of boom periods only. Conversely, in the case of point (b), it appears more plausible that the diffusion and swarming process is inhibited by the low levels of profitability that characterise depressions. In the innovation literature, support for the idea that depressions can both stimulate the development of basic innovations and shorten the lead times from one invention to the corresponding innovation is provided by Mensch (1979); whereas Freeman, Clark and Soete stress the favourable impact of booms on the swarming process and assert that clusters of basic innovations are not systematically related to depressions:

> The macro-economic effects of any basic innovation are scarcely perceptible in the first few years and often for much longer. What matters in terms of major economic effects is not the date of the basic innovation ...; what matters is the diffusion of this innovation – what Schumpeter vividly described as the 'swarming' process when imitators begin to realise the profitable potential of the new product or process and start to invest heavily. This swarming may not necessarily occur immediately after a basic innovation although it may do so if other conditions are favourable.[8] (Freeman, Clark and Soete, 1982, p. 65)

Problems connected with the collection and interpretation of data become even more serious as regards financial innovations. An interesting source is the data base developed by Silber (1983), which provides reliable figures on the most important innovations relating to instruments, markets and institutions in the period 1970–82. Silber adopts a literature-based counting procedure of the kind usually employed in the compilation of lists of technological innovations to gather information from trade publications and various financial journals.

I have followed an analogous procedure using Kindleberger's (1984) historical accounts for monetary events, financial events and milestones in the history of banking since the thirteenth century. This counting procedure is based, prevalently although not exclusively for European countries, on a comprehensive selection of novelties that were such for at least the country in which they were identified. Following Perez's and Tylecote's schemes summarised in Chapters 1 and 3, the adoption of the financial institutions and the occurrence of the events identified by Kindleberger can be associated with the more or less radical reform of the socio-institutional framework typical of upswings. As a first step in an empirical analysis of the long-term dynamics of technology and finance, Baker's and Kindleberger's data can therefore be used to construct a chart of the distribution of major patented innovations and major financial innovations and events over a very long period (Figures 5.1 and 5.2). For merely illustrative purposes – following a procedure widely employed in the long-waves literature (see Freeman, Clark and Soete, 1982; Kleinknecht, 1987) – Figures 5.1 and 5.2 present the original series of data and the same data transformed by an averaging process in order to flatten short-term fluctuations (which are not relevant to the present analysis) and emphasise long-term ones. The data plotted in these figures reveal that the distributions of both significant patents and significant financial innovations and events follow an irregular pattern. This pattern is even more irregular in the financial data.

However, owing to the employment of the averaging procedure,[9] these figures do not provide any statistically significant evidence for the existence of Kondratieff long waves as *true* cycles, since no demonstration of stationarity in variance and mean for the two series is forthcoming. It must be borne in mind, in effect, that although the averaging procedure removes the random component from fluctuations, it tends in turn to *create* cycles, in the sense that it imposes a systematic pattern upon the data that might not exist in the original series.

In order to detect whether the distribution of Baker's patent in the long wave conforms to Mensch's theory of 'bunching' of significant innovations, or to Freeman's 'random walk' hypothesis about basic innovations, I then computed the per annum growth rates (and their significance), the mean value, the variance and the skewness of the distribution of basic technological innovations for the various A and B periods in the twelve chronologies of long waves presented in Table 5.1. This simple procedure differs from those applied by Freeman, Clark and Soete (1982) and Kleinknecht to Baker's data,[10] and uses the original

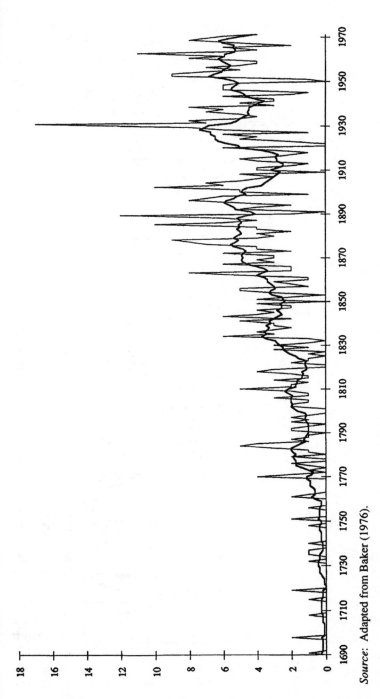

Source: Adapted from Baker (1976).

Figure 5.1 Most significant patented innovations according to Baker (absolute value and 10-year moving average)

70

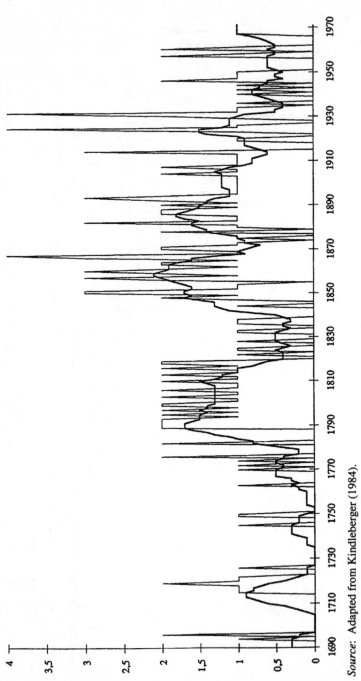

Source: Adapted from Kindleberger (1984).

Figure 5.2 Most significant financial innovations and events according to Kindleberger (absolute value and 10-year moving average)

series without further manipulation. Computation of such values in each subperiod serves to identify the extent to which the *pattern* of basic innovations conforms to observed cyclical fluctuations in some economic variables and remains positive or negative as economic growth approaches the upper and the lower turning point respectively. Moreover this analysis is helpful in identifying whether the *number* of basic innovations tends to diminish or increase in A and B periods of alleged Kondratieff long waves. The same procedure has been applied to the series of financial innovations and events. In this connection it is employed to detect whether these kinds of innovations and reforms are more likely to occur during boom periods, when they contribute to change the existing socio-institutional framework, than during depressions.

Growth rates have been computed employing least squares to estimate – for each A and B period identified in the various chronologies of long waves presented in Table 5.1 – the following function

$$\log Y_t = \beta_1 + \beta_2 \text{TREND} + u_t \tag{5.1}$$

where Y alternatively denotes the number of basic technological innovations and that of financial innovations, and TREND is a linear trend. Since the original innovation series comprises a number of zero values which would render problematic computation of their growth rates, all the observations have been augmented by 1 in order to permit computation of their natural logarithm. The estimated values of β_2 represent for each A and B period the annual average growth rate of the relevant distribution.

Annual growth rates, together with the value of the Student's t statistics, the mean, variance and the skewness of the two series in the various A and B periods are presented in Tables 5.2(a) and 5.2(b). Two main interpretations derive from these results.

As regards growth rates and their significance, in the case of basic technological innovations (Table 5.2(a)), none of the twelve chronologies displays a significant pattern within either the boom or depression periods of long waves. Accordingly, the hypothesis of the substantial independence of the *patterns* of basic innovations from cyclical fluctuations in other economic variables is confirmed.

On the other hand, the results are less consistent in using Kindleberger's financial data (Table 5.2(b)). In this case, if we look at the various couplings of A and B periods, analysis of two chronologies (De Wolff, Amin) shows that growth rates are higher and more significant during depressions than during booms.

Table 5.1   Long-wave chronologies according to different authors

Authors	1st Kondratieff		2nd Kondratieff		3rd Kondratieff		4th Kondratieff	
	Upswing	Down swing	Upswing	Down swing	Upswing	Down swing	Upswing	Down swing
Kondratieff	1790	1810/17	1844/51	1870/75	1890/96	1914/20	–	–
De Wolff	–	1825	1849/50	1873/74	1896	1913	–	–
Von Ciriacy-Wantrup	1792	1815	1842	1873	1895	1913	–	–
Schumpeter	1787	1813/14	1842/43	1869/70	1897/98	1924/25	–	–
Clark	–	–	1850	1875	1900	1929	–	–
Dupriez	1789/92	1808/14	1846/51	1872/73	1895/96	1920	1939/46	1974
Rostow	1790	1815	1848	1873	1896	1920	1935	1951
Mandel	–	1826	1847	1873	1893	1913	1939/48	1966
Van Duijn	–	–	1845	1872	1892	1929	1948	1973
Bouvier	–	–	1840	1865	1897	1913	–	–
Amin	1815	1840	1850	1870	1890	1914	1948	1967
Kuczynski	–	–	1850	1866	1896	1913	1951	1969

Source: Kleinknecht (1987).

Table 5.2(a) Average growth rates (g) and their significance (student's t distribution), mean value (x̄), variance (s²) and skewness (sk) of Baker's series of technological innovations, according to different long-wave chronologies

Long-wave chronologies – A and B periods

	A_1	B_1	A_2	B_2	A_3	B_3	A_4
Kondratieff	$g = 0.024$ (1.791)[d] $\bar{x} = 1.417$ $s^2 = 1.297$ $sk = 1.043$	$g = 0.025$ (2.954)[a] $\bar{x} = 2.257$ $s^2 = 2.726$ $sk = 0.475$	$g = 0.021$ (1.655)[e] $\bar{x} = 3.0$ $s^2 = 3.692$ $sk = 0.626$	$g = -0.011$ (−0.559)[f] $\bar{x} = 4.952$ $s^2 = 8.448$ $sk = 0.720$	$g = -0.024$ (−1.675)[e] $\bar{x} = 4.320$ $s^2 = 5.393$ $sk = 0.278$		
De Wolff		$g = 0.030$ (2.065)[c] $\bar{x} = 2.538$ $s^2 = 3.058$ $sk = 0.219$	$g = 0.030$ (2.056)[c] $\bar{x} = 3.080$ $s^2 = 3.910$ $sk = 0.515$	$g = -0.013$ (−0.755)[f] $\bar{x} = 5.087$ $s^2 = 8.083$ $sk = 0.620$	$g = -0.059$ (−2.403)[c] $\bar{x} = 4.611$ $s^2 = 6.722$ $sk = 0.043$		
Von Ciriacy-Wantrup	$g = 0.024$ (1.791)[d] $\bar{x} = 1.5$ $s^2 = 1.304$ $sk = 1.007$	$g = 0.028$ (2.247)[c] $\bar{x} = 2.0$ $s^2 = 2.519$ $sk = 0.590$	$g = 0.006$ (0.635)[f] $\bar{x} = 3.125$ $s^2 = 3.468$ $sk = 0.521$	$g = -0.013$ (−0.734)[f] $\bar{x} = 4.826$ $s^2 = 8.059$ $sk = 0.784$	$g = -0.053$ (−2.407)[c] $\bar{x} = 4.632$ $s^2 = 6.357$ $sk = 0.021$		
Schumpeter	$g = 0.019$ (1.770)[d] $\bar{x} = 1.464$ $s^2 = 1.221$ $sk = 0.961$	$g = 0.027$ (2.287)[f] $\bar{x} = 2.167$ $s^2 = 2.902$ $sk = 0.599$	$g = 0.004$ (0.279)[f] $\bar{x} = 3.107$ $s^2 = 3.581$ $sk = 0.484$	$g = 0.002$ (0.136)[f] $\bar{x} = 4.966$ $s^2 = 7.177$ $sk = 0.649$	$g = -0.029$ (−2.047)[c] $\bar{x} = 3.679$ $s^2 = 5.560$ $sk = 0.323$		

Table 5.2(a)   Continued

Long-wave chronologies – A and B periods

	A_1	B_1	A_2	B_2	A_3	B_3	A_4
Clark			$g = 0.029$ $(2.193)^c$ $\bar{x} = 3.115$ $s^2 = 3.786$ $sk = 0.471$	$g = -0.012$ $(-0.834)^f$ $\bar{x} = 5.077$ $s^2 = 7.914$ $sk = 0.474$	$g = -0.016$ $(-1.222)^f$ $\bar{x} = 3.9$ $s^2 = 5.955$ $sk = 0.222$		
Dupriez	$g = 0.022$ $(1.290)^f$ $\bar{x} = 1.476$ $s^2 = 1.362$ $sk = 1.043$	$g = 0.020$ $(2.613)^b$ $\bar{x} = 2.237$ $s^2 = 2.51$ $sk = 0.533$	$g = 0.022$ $(1.609)^e$ $\bar{x} = 3.038$ $s^2 = 3.788$ $sk = 0.572$	$g = -0.006$ $(-0.342)^f$ $\bar{x} = 4.958$ $s^2 = 8.129$ $sk = 0.666$	$g = -0.030$ $(-2.090)^c$ $\bar{x} = 4.280$ $s^2 = 5.793$ $sk = 0.234$	$g = 0.032$ $(1.707)^d$ $\bar{x} = 5.208$ $s^2 = 11.824$ $sk = 1.325$	$g = 0.018$ $(1.468)^e$ $\bar{x} = 5.379$ $s^2 = 6.458$ $sk = -0.186$
Rostow	$g = 0.023$ $(1.959)^d$ $\bar{x} = 1.462$ $s^2 = 1.298$ $sk = 0.949$	$g = 0.028$ $(3.151)^a$ $\bar{x} = 2.265$ $s^2 = 2.746$ $sk = 0.481$	$g = 0.022$ $(1.609)^e$ $\bar{x} = 3.038$ $s^2 = 3.798$ $sk = 0.572$	$g = -0.006$ $(-0.342)^f$ $\bar{x} = 4.958$ $s^2 = 8.129$ $sk = 0.666$	$g = -0.030$ $(-2.090)^c$ $\bar{x} = 4.280$ $s^2 = 5.793$ $sk = 0.234$	$g = 0.086$ $(2.290)^c$ $\bar{x} = 5.250$ $s^2 = 16.333$ $sk = 1.199$	$g = -0.060$ $(-2.255)^c$ $\bar{x} = 4.294$ $s^2 = 5.471$ $sk = -0.331$
Mandel		$g = 0.046$ $(2.973)^a$ $\bar{x} = 2.727$ $s^2 = 2.874$ $sk = 0.241$	$g = 0.021$ $(1.655)^e$ $\bar{x} = 3.0$ $s^2 = 3.692$ $sk = 0.626$	$g = -0.011$ $(-0.559)^f$ $\bar{x} = 4.952$ $s^2 = 8.448$ $sk = 0.720$	$g = -0.034$ $(-1.712)^d$ $\bar{x} = 4.476$ $s^2 = 6.062$ $sk = 0.142$	$g = 0.024$ $(2.047)^c$ $\bar{x} = 4.562$ $s^2 = 10.512$ $sk = 1.554$	$g = 0.025$ $(1.375)^e$ $\bar{x} = 5.217$ $s^2 = 7.632$ $sk = -0.055$

Table 5.2(a)   Continued

Long-wave chronologies – A and B periods

	A_1	B_1	A_2	B_2	A_3	B_3	A_4
Van Duijn			$g = 0.017$ $(1.401)^a$ $\bar{x} = 3.071$ $s^2 = 3.55$ $sk = 0.542$	$g = -0.019$ $(-0.904)^f$ $\bar{x} = 5.048$ $s^2 = 8.448$ $sk = 0.630$	$g = -0.014$ $(-1.592)^c$ $\bar{x} = 4.079$ $s^2 = 5.858$ $sk = 0.111$	$g = -0.041$ $(-2.561)^b$ $\bar{x} = 5.800$ $s^2 = 11.955$ $sk = 1.645$	$g = 0.017$ $(1.045)^f$ $\bar{x} = 5.625$ $s^2 = 6.418$ $sk = -0.194$
Bouvier			$g = -0.009$ $(-0.644)^f$ $\bar{x} = 3.038$ $s^2 = 3.638$ $sk = 0.513$	$g = 0.007$ $(0.766)^f$ $\bar{x} = 4.697$ $s^2 = 6.905$ $sk = 0.801$	$g = -0.056$ $(-2.048)^d$ $\bar{x} = 4.412$ $s^2 = 6.382$ $sk = 0.130$		
Amin		$g = -0.060$ $(-2.255)^d$ $\bar{x} = 4.294$ $s^2 = 5.471$ $sk = -0.331$	$g = 0.046$ $(2.973)^d$ $\bar{x} = 2.727$ $s^2 = 2.874$ $sk = 0.241$	$g = 0.021$ $(1.655)^f$ $\bar{x} = 3.0$ $s^2 = 3.692$ $sk = 0.626$	$g = -0.056$ $(-2.048)^f$ $\bar{x} = 4.412$ $s^2 = 6.382$ $sk = 0.130$	$g = 0.007$ $(0.766)^c$ $\bar{x} = 4.697$ $s^2 = 6.905$ $sk = 0.801$	$g = 0.017$ $(1.045)^f$ $\bar{x} = 5.625$ $s^2 = 6.418$ $sk = -0.194$
Kuczynski	$g = 0.007$ $(0.766)^d$ $\bar{x} = 4.697$ $s^2 = 6.905$ $sk = 0.801$		$g = 0.038$ $(1.380)^e$ $\bar{x} = 2.706$ $s^2 = 4.471$ $sk = 0.812$	$g = 0.002$ $(0.214)^f$ $\bar{x} = 4.710$ $s^2 = 6.946$ $sk = 0.844$	$g = -0.056$ $(-2.403)^c$ $\bar{x} = 4.611$ $s^2 = 6.772$ $sk = 0.043$	$g = 0.006$ $(0.672)^f$ $\bar{x} = 4.385$ $s^2 = 9.927$ $sk = 1.410$	$g = 0.006$ $(0.343)^f$ $\bar{x} = 6.0$ $s^2 = 6.111$ $sk = -0.063$

Notes: (a) significant at 0.005; (b) significant at 0.01; (c) significant at 0.025; (d) significant at 0.05; (e) significant at 0.10; (f) not significant.

Table 5.2(b) Average growth rates (g) and their significance (Student's t distribution), mean value ($\bar{x}$), variance (s^2) and skewness (sk) of Kindleberger's series of financial innovations and events, according to different long-wave chronologies

	Long-wave chronologies – A and B periods						
	A_1	B_1	A_2	B_2	A_3	B_3	A_4
Kondratieff	$g = -0.010$ $(-1.747)^d$ $\bar{x} = 2.5$ $s^2 = 0.261$ $sk = -0.000$	$g = -0.018$ $(-2.879)^a$ $\bar{x} = 1.629$ $s^2 = 0.476$ $sk = 0.593$	$g = 0.002$ $(0.251)^f$ $\bar{x} = 2.741$ $s^2 = 0.892$ $sk = 0.506$	$g = 0.044$ $(3.440)^a$ $\bar{x} = 2.286$ $s^2 = 0.814$ $sk = 0.230$	$g = -0.022$ $(-2.579)^b$ $\bar{x} = 2.160$ $s^2 = 0.557$ $sk = 0.913$		
De Wolff		$g = 0.017$ $(1.562)^e$ $\bar{x} = 1.577$ $s^2 = 0.574$ $sk = 1.343$	$g = -0.010$ $(-0.869)^f$ $\bar{x} = 2.720$ $s^2 = 1.043$ $sk = 0.322$	$g = 0.035$ $(3.068)^b$ $\bar{x} = 1.304$ $s^2 = 0.767$ $sk = 0.191$	$g = 0.002$ $(0.273)^f$ $\bar{x} = 2.111$ $s^2 = 0.105$ $sk = 2.272$		
Von Ciriacy-Wantrup	$g = -0.010$ $(-1.638)^e$ $\bar{x} = 1.417$ $s^2 = 0.254$ $sk = 0.317$	$g = -0.022$ $(-2.474)^b$ $\bar{x} = 0.607$ $s^2 = 0.470$ $sk = 0.626$	$g = 0.021$ $(2.695)^b$ $\bar{x} = 1.500$ $s^2 = 1.097$ $sk = 0.408$	$g = 0.036$ $(3.258)^b$ $\bar{x} = 1.304$ $s^2 = 0.767$ $sk = 0.181$	$g = 0.002$ $(0.389)^f$ $\bar{x} = 1.105$ $s^2 = 0.099$ $sk = 2.372$		
Schumpeter	$g = 0.004$ $(0.483)^f$ $\bar{x} = 1.393$ $s^2 = 0.396$ $sk = -0.471$	$g = -0.022$ $(-2.891)^a$ $\bar{x} = 0.600$ $s^2 = 0.455$ $sk = 0.625$	$g = 0.026$ $(2.648)^b$ $\bar{x} = 1.571$ $s^2 = 1.143$ $sk = 0.347$	$g = 0.013$ $(1.504)^e$ $\bar{x} = 1.310$ $s^2 = 0.650$ $sk = 0.209$	$g = -0.011$ $(-1.128)^f$ $\bar{x} = 1.036$ $s^2 = 0.776$ $sk = 1.501$		

Table 5.2(b)   Continued

Long-wave chronologies – A and B periods

	A_1	B_1	A_2	B_2	A_3	B_3	A_4
Clark			$g = -0.011$ $(-1.026)^f$ $\bar{x} = 1.692$ $s^2 = 1.022$ $sk = 0.385$	$g = 0.015$ $(1.627)^b$ $\bar{x} = 1.308$ $s^2 = 0.622$ $sk = 0.385$	$g = -0.014$ $(-1.658)^f$ $\bar{x} = 0.967$ $s^2 = 0.792$ $sk = 1.480$		
Dupriez	$g = -0.013$ $(-1.779)^d$ $\bar{x} = 1.476$ $s^2 = 0.262$ $sk = 0.089$	$g = -0.014$ $(-2.336)^c$ $\bar{x} = 0.684$ $s^2 = 0.492$ $sk = 0.491$	$g = 0.0001$ $(0.014)^f$ $\bar{x} = 1.769$ $s^2 = 0.305$ $sk = 0.446$	$g = 0.031$ $(2.951)^e$ $\bar{x} = 1.292$ $s^2 = 0.739$ $sk = 0.235$	$g = -0.016$ $(-1.874)^d$ $\bar{x} = 1.0$ $s^2 = 0.417$ $sk = 0.892$	$g = -0.017$ $(-1.131)^f$ $\bar{x} = 0.875$ $s^2 = 1.245$ $sk = 1.674$	$g = -0.001$ $(-0.171)^f$ $\bar{x} = 0.594$ $s^2 = 0.443$ $sj = 0.623$
Rostow	$g = -0.011$ $(-2.279)^c$ $\bar{x} = 1.462$ $s^2 = 0.258$ $sk = 0.145$	$g = -0.011$ $(-1.508)^f$ $\bar{x} = 0.618$ $s^2 = 0.486$ $sk = 0.633$	$g = 0.0001$ $(0.014)^f$ $\bar{x} = 1.769$ $s^2 = 0.905$ $sk = 0.446$	$g = 0.031$ $(2.951)^e$ $\bar{x} = 1.292$ $s^2 = 0.737$ $sk = 0.235$	$g = -0.013$ $(-1.874)^f$ $\bar{x} = 1.0$ $s^2 = 0.417$ $sk = 0.892$	$g = 0.012$ $(0.414)^f$ $\bar{x} = 1.125$ $s^2 = 1.583$ $sk = 1.288$	$g = 0.028$ $(1.528)^e$ $\bar{x} = 0.588$ $s^2 = 0.382$ $sk = 0.434$
Mandel		$g = 0.002$ $(0.200)^f$ $\bar{x} = 0.364$ $s^2 = 0.242$ $sk = 0.529$	$g = 0.002$ $(0.251)^f$ $\bar{x} = 1.741$ $s^2 = 0.892$ $sk = 0.506$	$g = 0.044$ $(3.440)^a$ $\bar{x} = 1.286$ $s^2 = 0.814$ $sk = 0.230$	$g = -0.011$ $(-1.599)^e$ $\bar{x} = 1.238$ $s^2 = 0.290$ $sk = 1.997$	$g = -0.004$ $(-0.377)^f$ $\bar{x} = 0.844$ $s^2 = 1.168$ $sk = 1.636$	$g = -0.021$ $(-1.632)^e$ $\bar{x} = 0.565$ $s^2 = 0.530$ $sk = 0.800$

Table 5.2(b)   Continued

Long-wave chronologies – A and B periods

	A_1	B_1	A_2	B_2	A_3	B_3	A_4
Van Duijn			$g = 0.018$ $(1.838)^d$ $\bar{x} = 1.643$ $s^2 = 1.053$ $sk = 0.322$	$g = 0.031$ $(2.262)^c$ $\bar{x} = 1.190$ $s^2 = 0.662$ $sk = 0.203$	$g = -0.016$ $(-2.785)^a$ $\bar{x} = 1.079$ $s^2 = 0.777$ $sk = 1.235$	$g = -0.020$ $(-1.114)^f$ $\bar{x} = 0.900$ $s^2 = 0.937$ $sk = 1.509$	$g = 0.006$ $(0.544)^f$ $\bar{x} = 0.577$ $s^2 = 0.414$ $sk = 0.592$
Bouvier			$g = 0.040$ $(3.794)^a$ $\bar{x} = 1.269$ $s^2 = 1.085$ $sk = 0.293$	$g = 0.003$ $(0.412)^f$ $\bar{x} = 1.424$ $s^2 = 0.877$ $sk = 0.656$	$g = 0.001$ $(0.144)^f$ $\bar{x} = 1.118$ $s^2 = 0.110$ $sk = 2.167$		
Amin	$g = -0.020$ $(-1.943)^d$ $\bar{x} = 0.654$ $s^2 = 0.475$ $sk = 0.519$	$g = 0.124$ $(3.973)^a$ $\bar{x} = 0.727$ $s^2 = 1.018$ $sk = 1.022$	$g = 0.005$ $(0.375)^f$ $\bar{x} = 1.857$ $s^2 = 1.029$ $sk = 0.268$	$g = 0.020$ $(1.324)^f$ $\bar{x} = 1.238$ $s^2 = 0.680$ $sk = 0.071$	$g = -0.006$ $(-0.854)^f$ $\bar{x} = 1.360$ $s^2 = 0.407$ $sk = 1.433$	$g = 0.001$ $(0.115)^f$ $\bar{x} = 0.857$ $s^2 = 1.126$ $sk = 1.566$	$g = -0.009$ $(-0.561)^f$ $\bar{x} = 0.500$ $s^2 = 0.474$ $sk = 0.920$
Kuczynski			$g = -0.002$ $(-0.107)^f$ $\bar{x} = 1.824$ $s^2 = 0.804$ $sk = -0.085$	$g = 0.004$ $(0.439)^f$ $\bar{x} = 1.452$ $s^2 = 0.923$ $sk = 0.570$	$g = 0.002$ $(0.273)^f$ $\bar{x} = 1.111$ $s^2 = 0.105$ $sk = 2.272$	$g = -0.001$ $(-0.079)^f$ $\bar{x} = 0.846$ $s^2 = 1.028$ $sk = 1.627$	$g = 0.022$ $(1.294)^f$ $\bar{x} = 0.474$ $s^2 = 0.485$ $sk = 1.022$

Notes: (a) significant at 0.005; (b) significant at 0.01; (c) significant at 0.025; (d) significant at 0.05; (e) significant at 0.10; (f) not significant

The most interesting results arise from a comparison of mean values during successive A and B periods in the various chronologies. Using a standard *t*-test for the differences between means, it emerges that the difference between the higher (lower) mean value for technology data (finance data) during depressions and the lower (higher) mean value during booms is statistically significant, in particular for the second long wave (first long wave). These findings suggest that:

1. The financial component of the overall socio-institutional framework is more likely to be changed during upswings than downswings.
2. The number of basic technological innovations tends to be higher during depression periods – a point that can be directly traced back to Schumpeter (see Section 2.3 above) – when the profit motive sets in motion a trial-and-error search process exploring what would be potentially economically profitable (see Perez, 1985, p. 43).

As regards variance alone, the results in the case of financial innovations are rather inconclusive, whereas variance is particularly high in those of basic technological innovations, and more so in B than in A periods. This finding supports the hypothesis advanced in Section 3.3 to explain the search process during phases characterised by the obsolescence of active technological styles, and it confirms the conclusions drawn from analysis of the mean value. In effect, the result shows that such innovations tend to be more concentrated during depression periods – when potential basic innovations are more likely to be experimented with, and in some cases rejected – although no decisive evidence is provided that they are always less concentrated during boom periods.[11]

According to these results – although it is true that diffusion clusters are more important than basic innovations in terms of their impact on economic growth and structural change – it appears likely that during depressions the search for technological innovations also moves in directions that differ from those entailed by the active – but obsolescent and rapidly approaching exhaustion – technological style.

5.3 THE INTERACTION OF FINANCIAL AND TECHNOLOGICAL STYLES IN KONDRATIEFF LONG WAVES: A HISTORICAL ANALYSIS

The test carried out in the previous section on a group of industrialised countries revealed historical differences (a) only in the mean value and the

variance in the distribution of *basic* technological innovations if at all, and
(b) as regards the mean value of the distribution of important financial
innovations and events, only for certain couplings of upswing and
downswing periods that match the periodisation of Kondratieff long
waves. However – due to the shortness of the time series and the
inevitable crudity of the statistical analysis – the above results say nothing
about whether these differences result from historically unique events or
whether they are likely to be repeated in the future. Therefore this section
conducts a 'qualitative' investigation of the interactions between changes
in the *technological style* and changes in that particular aspect of the
socio-institutional framework represented by the *financial style* during five
alleged Kondratieff long waves. The object of analysis in this case is the
swarming process, and the way in which a new techno-economic
paradigm takes shape around a new leading technological style as a result
of, or simultaneously with, a change in the socio-institutional framework.
Again following the argument originally propounded by Perez (1985), I
maintain that:

> it is possible to identify each successive Kondratieff wave with the
> deployment of a specific, all-pervasive, technological revolution. In
> other words, that behind the apparently infinite variety of technologies
> of each long-wave upswing, there is a distinct set of accepted 'common-
> sense' principles, which define a broad technological trajectory towards
> a general 'best-practice' frontier. These principles are applied in the
> generation of innovations and in the organization of production in one
> firm after another, in one branch after another, within and across
> countries. (Perez, 1985, p. 43)

In this connection, I suggest the couplings between *active* technological
styles (each of which represents a diffusion cluster) and financial styles set
out in Table 5.3.

In this simple scheme, each technological style or group of technological
styles is associated with a financial style, according to the definitions given
in Section 3.3. Singled out for each alleged Kondratieff long wave is
the most significant financial institution(s), event(s), or institutional
arrangement(s), which denotes the emergence of a new financial style and
accounts for a radical restructuring of the socio-institutional framework.
Accordingly, what constitutes a financial style is not necessarily a cluster
of financial innovations, but a single innovation or a well-defined group of
innovations and events that actually cause the financial framework to be
restructured.

Table 5.3   Couplings of technological and financial styles during five alleged
Kondratieff long waves

Long waves	Technological styles	Financial styles
I	Specialised machinery for the textile industry ('water style')	Use of merchant capital in industrial activities
II	Railroading and transportation (exploiting *motive* steam power)	Stock exchange market
III	Chemicals, electricity, steel	International gold standard, Universal Bank
IV	Internal combustion engine, petroleum refining ('Fordist' style)	Central Bank control function
V	Information technology, biotechnology	Non-banking financial institutions (venture capital, pension funds, etc.)

The analysis carried out in this section as regards both financial and technological innovations is transnational. It mainly concerns the four earlier developed countries (France, Germany, UK, US) plus Japan and Italy, although other countries are also taken into consideration.

5.3.1   The First Industrial Revolution

The first industrial revolution was almost entirely confined to the UK, at a time when there were no other industrialised countries in the world economy. During this period, therefore, Britain was the leader country in the production and exportation of the most technologically advanced products.

The rate of diffusion of technological progress and capital formation accelerated rapidly in the UK during the first industrial revolution (see Kuznets, 1971): the widespread adoption of water wheels as an efficient source of power, along with faster water transport, characterised the upswing technologically (see Tylecote, 1993, pp. 39–45). However the most significant technological innovations took place in the field of capital equipment, starting with Richard Arkwright's water frame method of spinning[12] and culminating, after the original inventions by Thomas Savery in 1698, Thomas Newcomen in 1712 and James Watt in 1769, in the diffusion of steam power as the principal source of power for industrial equipment.

In fact, from a technological perspective the manufacturing system underwent profound transformations in production techniques consequent on the introduction of new and more effective machinery, which played a major role in the dematuring of the textile and clothing industries (see von Tunzelmann, 1978). This dematurity process took the form of new standards in product design and the organisation of production achieved by abandoning previous routines and by introducing new skills, expertise and knowledge. This transformation occurred within an industry that had been previously typified by slow technological change and affected by a trade-off in the exploitation of navigable rivers for either transport or power production. As Tylecote (1993, p. 38) observes, increased demand for water power led to a proliferation of weirs in the navigable rivers, which, by obstructing navigation, made the use of the waterway system for the transportation of intermediate and consumer goods inefficient and costly. The situation changed during the eighteenth century: the availability of more technologically advanced water wheels and increasing recourse to 'static' steam power instead of water power (see Scherer, 1984) enhanced the efficiency and lowered the costs of using the waterway system, thereby inducing significant productivity gains and reducing delivery times. None the less, the British Industrial Revolution was not characterised by widespread employment of new materials nor did capital intensity and firm size increase very significantly (Pollard, 1964).

According to the chronologies presented in Table 5.1, the upswing of this first Kondratieff long wave started around 1790. Therefore the advances in spinning and power production technologies mentioned above were followed by a long period of inertia due to resistance to change in the socio-institutional framework. Before these technological innovations could be fully adopted and generate the emergence of a new techno-economic subsystem, the socio-institutional framework had to undergo radical restructuring. As regards the component of the socio-institutional framework represented by financial institutions and arrangements, however, one distinct major change is difficult to specify. Economic historians usually identify a series of main capital sources issued by several different economic sectors in this period (see Crouzet, 1972, p. 164), so that it is only possible to provide an arbitrary ordinal sequence for the most important sources of capital. The most significant distinction for our present purposes is between capital arising within industry itself (internal financing) and capital previously accumulated in the commercial sector but playing a major role in financing industrial activities (external financing).

In the case of internal sources, start-up capital sometimes moved from lower to upper organisational levels, or from artisan to small manufacturing firms. Frequently, manufacturing industry used capital accumulated in the commercial sector as an external source of funds although, owing to relatively low barriers to entry and initial capital expenditures, recourse to self-financing was in most cases preferred. However, as fixed capital requirements increased, the importance of self-financing tended to decrease at the core of the expansionary phase. This was to a large extent the effect of technical progress and industrial concentration, which raised the threshold of entry into several industrial sectors. Would-be entrepreneurs therefore became increasingly familiar with external financing, frequently in the form of personal loans. Thus lending was not yet an institutionalised activity and was mostly carried out by rich merchants who had accumulated their wealth in previous years and were frequently active in the same sector as the one to which they lent their money. Nevertheless lending performed a central function in the promotion of industrial activities. The phenomenon may be seen as a primitive form of venture capital, since not only did the merchant–financier provide the borrower with risk capital, he also acted as a consultant, exploiting his knowledge of the market that the new industrial firm was trying to enter. Only an informal financial structure was possible in the segmented economic system of the time. The circulation of information was still hindered by environmental factors, with the result that the financial market had only local relevance. This largely 'unofficial' financial sector therefore had a major role to play via intermediation, although primarily *within* industrial networks.

During the first industrial revolution the role of the banking sector in providing financial support to innovative industrial activities seems instead to have been of little significance. Until the end of the eighteenth century the European banking system consisted mostly of public banks – which acted in the same way as the medieval money-changers – and private bankers, who, like their medieval counterparts, were mainly exchange dealers (see De Roover, 1954). Hence none of these institutions contributed greatly to the financing of investment activities in the fledgeling techno-economic subsystem, and this may partially explain why the requirements of financial capital remained relatively small during the whole of the first industrial revolution.

In conclusion, although representing a significant progress with respect to previous ones, the technologies adopted for the early years of British industrialisation used little fixed capital. This was due to a two-way link between the kinds of capital funding available (principally merchants, in the form of working capital) and the nature of the technologies used (employing local raw materials, little plant and equipment, etc.).

5.3.2   The Mid-Nineteenth-Century Long Wave

During the alleged second Kondratieff long wave – that is, the development cycle centred on the 1850s – some of the leading industries to have emerged in the first industrial revolution went into relative decline, with the concomitant rapid development of new, more capital-intensive industries concentrated in large urban areas. The most significant technological events of this period occurred in transportation (both by land and by sea), and it was in particular the railways that altered the scale of fixed capital requirements and contributed greatly to economic expansion. As transport by land grew increasingly cheap, the cost of water power and that of steam power steadily declined. The outcome was a dramatic fall in the cost of most of the key production factors and the lower effective cost of industrial energy (Tylecote, 1993, p. 47).

Although Britain was clearly the initiator of this revolution in transport, during the second Kondratieff long wave America too entered the technological arena, although mainly as a borrower of the railways style (see Rosenberg, 1972).[13] In fact, in Britain inventions associated with the railways style had been under development since the end of the eighteenth century. In 1804, by applying inventions developed for steamboats during the 1790s by various individual inventors, Richard Trevithick successfully ran the first steam locomotive on a prepared track at Penydarran in Cornwall. In 1805–6 a Mr Blackett ordered a locomotive from Trevithick, but it was never used because it was too heavy for existing rails. It was therefore only in 1814 that the first steam locomotive – George Stephenson's successful development of early prototypes – was used to transport coal from the Tyneside collieries to the river, where it was loaded on steamboats (see Jewkes *et al.*, 1962). Finally, by 1816 several locomotives were in operation in the UK, thanks to a new and stronger kind of rail. It was, however, only in the early 1830s that American industry began to build locomotives, thereby giving a transnational nature to the newly emerging railways style.

Accepting the periodisations presented in Table 5.1, the 'Railways Kondratieff' also displays a mismatch between major changes in technology and their assimilation in the techno-economic subsystem. As in the case of the first industrial revolution, the relatively long period of inertia can be explained by the resistance raised by the existing socio-institutional framework. However, for the second Kondratieff long wave it is much easier than it was in the case of the first industrial revolution to identify the crucial reform that occurred in the financial component of the socio-institutional framework. The 'railways revolution' interacted with the socio-institutional revolution brought about by the increasing importance of the stock exchange market. In Britain, 'railways greatly increased trading in non-government debt in London, and led to the development of stock exchange in provincial towns' (Hawke, 1970, p. 390). But also in the US, during a period in which technology became increasingly scale-dependent (see Chandler, 1965), the stock exchange was the main source of capital for railroading investment. The transnational character assumed by the stock exchange financial style during this period reflected the need to mobilise a huge amount of capital to finance individual firms obliged to operate on a very large scale to be successful.

As Crouzet (1972) has shown, the level of investment in railroading was particularly high in Britain as the new technological style spread, especially during the peak years between 1845 and 1850. This finding is confirmed by the dramatic increase during those years in the ratio of gross fixed investment in the railroading sector to gross national product (Figure 5.3). It would therefore be incorrect to interpret the direction of the causal linkage as running from the railways to the stock exchange. According to Mitchell (1964, p. 333) 'the introduction of Railways in Britain did not have a very great immediate impact on the economy, the main effects not being felt until during and after the great mania of

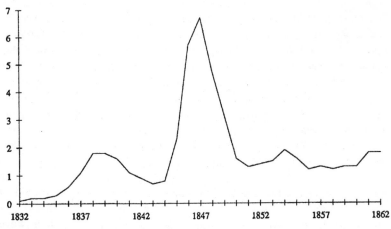

Source: Adapted from Mitchell (1964).

Figure 5.3   Gross fixed investments in the railroading sector as a share of gross national product (UK, 1832–62)

the 1840s'. As we know, that 'great mania' consisted of playing the stock exchange.

The period of stock exchange resurgence followed years marked by recurrent banking panics both in Britain and the USA (see McGrane, 1924). During the decades prior to the stock exchange boom, the value of every kind of property had risen sharply, and a substantial redistribution of income among different social classes had taken place, especially in the USA.

As Mc Grane (1924, p. 91) observes 'a restless spirit of adventure and daring enterprise swept the nation. Over-trading, speculation, and investments in unproductive undertakings became the dominant note in American society'. This situation quickly imploded – probably after the banks suspended specie payments – and a collapse in credit ensued. The succession of banking crises forced money holders to invest their assets in different ways. It was therefore no coincidence that the stock exchange grew so rapidly in the 1840s; and the main part of this expansion, in preference shares, was brought about by railroading companies. A new style of employing savings emerged, not only among businessmen but also among merchants and ordinary people: the interaction between this new dynamic style of investing and the new technological style based on the exploitation of *motive* steam power permitted the new techno-economic subsystem to realise its full growth potential.

The widespread diffusion of the railways technological style and the stock exchange financial style during this period was associated with other significant changes. In the socio-institutional framework, mention must of course be made of the emergence of the bourgeoisie in financial activities, which was the most innovative social phenomenon of the 1840s. The new middle class was forced by the unprecedented conditions in the financial markets to play the stock exchange instead of investing its money in more traditional ways, or 'hiding savings under the mattress'. A significant organisational innovation in the railroading sector was the introduction of the limited liability public corporation, the shares of which could be marketed in relatively small quantities to a broad public. In the specific case of railroading companies, new shares were marketed by investment bankers through the stock exchange market.

5.3.3   The Second Industrial Revolution

The third long wave was characterised by the fast technological development of Germany and the USA, which eventually overtook Britain in technological innovation, and in the exportation of most industrial production.

The main technological features of this period, which has been called the 'second industrial revolution', were the prevalence of product over process innovations and the widespread development of heavy industries. New industries emerged that exploited machine-based technologies to manufacture entirely (or at least partially) new products, such as automobiles, telephones, chemicals, fluorescent lamps and so on. Owing to the abundance of important novelties in most industrial sectors, during this third Kondratieff long wave it is difficult to identify one leading technological style. However it is conceivable that at least three different *active* styles significantly affected the evolution of the techno-economic subsystem: chemicals, electricity and steel.

As regards chemicals, Mensch (1979) has identified the formation of an innovative cluster during the third alleged Kondratieff upswing, when the development of synthetic dyes enabled Germany to challenge, and eventually to take the leadership from, Britain in the industry as a whole (Walsh, 1984). This evolution in chemicals was stimulated by the early establishment of independent and industrial R&D laboratories coupled with 'a tremendous investment in both academic and industrial research and development, the training of large numbers of scientists and engineers by establishing scientific institutions, and various official policies relating to finance and patents' (Walsh, 1984, p. 218). In this connection, the restructuring of the chemicals industry during the third Kondratieff long wave is an example of the transition from 'heroic' to 'trustified' capitalism: science and technology became endogenous economic factors and large firms acquired a dominant position in innovation, as the supremacy of the German dye industry at the turn of the nineteenth century clearly shows.

The electricity style was largely responsible for determining the general patterns of process innovation by providing a cheaper source of power, which permitted the introduction of the first production lines. In fact Devine's (1983) chronology of electrification in industry shows that the innovative cluster was formed in the two decades preceding the turn of the century.

Significant innovations in steel production brought about the emergence of a steel technological style, which generated new demand and offered new opportunities for engineering to expand. The first step in the transition from iron to steel was, in 1854, the Bessemer converter which, by blowing air through molten pig iron, removed some of the impurities (carbon and silicon) from it to produce steel. These impurities, together with phosphorous, restricted mass production of steel, and in this connection the Bessemer process therefore represented an important breakthrough (see Rosenberg, 1985). However it was only with the introduction, in 1878–9, of Sidney Gilchrist-Thomas's 'basic' process for removing phosphorous from steel that it became possible to 'produce steel to the increasingly exacting specifications required for many of the new machines and products of the twentieth century' (Rosenberg, 1982, p. 91). The availability of higher quality steel at much lower prices revolutionised the engineering industry, and permitted new or better-made steel-intensive equipment to be brought to the market.

In the course of the third Kondratieff long wave, the complementarities between old and new technological styles also grew increasingly close, giving rise to an unprecedented dematurity process of the kind described in Section 3.3. A significant example can be provided by looking at Schmookler's (1972) analysis of patent data. Schmookler's figures reveal that the employment of electricity in locomotives triggered a dematurity process in the railways style, with the formation of an innovative cluster around the 1890s.

As in previous long waves, in the third one, too, the emergence of a new financial style was necessary to reduce resistance to change and enable the new technological style to spread. The second industrial revolution was characterised in this respect by the adoption of the gold standard (Maddison, 1977). Indeed, in the twenty-five years between 1890 and 1914 gold was the ultimate and *effective* numeraire in most countries, and other means of payment were readily redeemable in gold at their bearer's request (De Cecco, 1974). The international gold standard may therefore be identified as the predominant financial style of the third Kondratieff upswing, and it greatly favoured the growth of international trade and the flows of capital among countries, thereby permitting the assimilation of new technologies in the techno-economic subsystem. According to De Cecco:

> This was the time ... when a very large number of countries attempted to get on the gold standard. The Governments of countries as diverse as [Britain, US, Belgium, Italy, Japan, Switzerland, The Netherlands, the Scandinavian countries] Germany, Austria, Hungary and Argentina all tried their hand at monetary reform, and all in the pursuit of a common objective, which was sometimes reached, but more often missed.... It was during this period that most of the monetary reforms took place whose general aim was the adoption of the gold standard. (De Cecco, 1974, p. 40)

In all countries, the decision to place their monetary systems on the gold standard stemmed from the belief that industrialisation would be stimulated by the intervention of foreign capital. Monetary stability was therefore crucial for the capital-receiving countries, which had to attract foreign capital and repatriate principal and interest without losses.

In the financial systems of at least three of the countries – Germany, Italy and Japan – which adopted the gold standard, an important role in intermediation was played by universal banks providing *capital credit*. This type of credit greatly helped to finance the investment outlays of industrial firms, and thus gave rise to long-term relationships in which the bank was the dominant partner. Generally speaking, when this particular kind of financing is sought, the bank holds currency that can be easily converted into monetary capital, just as the firm depends on its own machinery and capital equipment to transform goods into money. When the currency is converted into equities, the bank has representatives sitting on the firm's board of directors. It can exert the voting function on behalf of its clients, who deposit their voting shares in the bank facilities (*Depotstimmrecht* in Germany). The universal bank usually has a closer relationship with large corporations than with small and medium-sized firms. This linkage indirectly increases the risk of financial instability. In fact, if a large corporation controlled by a bank of this type goes bankrupt, the bank itself will probably be unable to cover the losses caused by the firm's bankruptcy, thereby undermining the stability of the whole financial system.

Universal banks were a product of the 1860s and 1870s in continental Europe (e.g. the Crédit Lyonnais and the Deutsche Bank), although there were antecedents from the late eighteenth century like the French and Russian land (mortgage) banks, which however did not undertake ordinary deposit banking. In Germany, the universal bank became of particular importance during the third alleged Kondratieff long wave, given its ability to satisfy the financial requirements of large companies.[14] The passage from an agro-industrial production model to a modern industrial system was typified in Germany by the emergence of large companies with higher financial requirements. The universal bank therefore performed a crucial function in channelling financial resources towards such companies (see Tilly, 1986).

In Italy this financial institution lost its appeal when the linkages between banks and the armaments industry became especially close; in some cases with disastrous consequences for both organisations. A case in point was the bankruptcy of Banca Italiana di Sconto consequent on its involvement in the financing of Società Ansaldo,[15] an event that brought about a serious banking crisis with deleterious effects on both the stock exchange market and the national currency.

In Japan the universal bank has remained at the core of the financial system. Its modern version is represented by the so-called *main bank* (to which I return in Section 7.2.3) which has financed most of Japanese technological development over the last twenty years.

5.3.4 The Fourth Alleged Kondratieff Long Wave

Although this is not the most appropriate place to discuss the various features and implications of the widespread diffusion of the internal combustion engine, it must be recalled that not only did this technology provide the basis for the 'Fordist style' characteristic of the fourth alleged Kondratieff long wave, but it is probably one of the most long-lasting technological styles to have appeared since the end of the nineteenth century (Abernathy, Clark and Kantrow, 1983).

By the early decades of this century a tendency emerged within the automobile industry towards standardisation and mass production, of which the Model T Ford is the classic example. This tendency towards mass production and, consequently, industrial concentration can be considered to be the main structural feature of this period: process innovations brought about by the introduction of new machinery and capital equipment characterised the fourth Kondratieff upswing, while the scale of R&D became so vast that collaborative ventures had to be undertaken. Moreover, the restructuring process that took place in manufacturing was to a large extent dictated by the more efficient exploitation of petroleum and its derivatives made possible by the emergence of an 'oil technological style'.

The most important inventions in petroleum refining were made during the 1930s (see Figure 5.4) with the discovery of the batch thermal process in 1931 and of catalytic cracking in 1936 (see Enos, 1962). The batch thermal process was a further development of an already existing process that obtained kerosene from crude oil at atmospheric pressure. The new process was based on the discovery that by increasing the pressure it was possible to obtain petrol from crude oil. The cracking process instead involves a treatment – called 'catalysis' – of hydrocarbons for heat exchange. Developed by a French inventor, Eugene Houdry, who had begun research into the nature of catalysis at the end of the First World War, cracking is still the main technique employed in petrol production.[16]

As emerges from Figure 5.4, an increasing number of patents relating to petroleum refining were issued during the 1930s, in parallel with major improvements and inventions in aircraft style. For example in 1936 the development of the DC-3 aircraft revolutionised the design and production

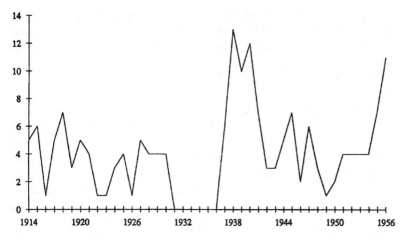

Source: Adapted from Schmookler (1966).

Figure 5.4 Important inventions in the petroleum industry (all countries, 1914–56)

of airplanes, while in 1937 the issue of the first patent for the turbo-jet engine ushered in a new era in transportation.

Quite different was the case of electronics (see Nelson, 1962), which during this period for the first time assumed the features of a latent technological style (see Section 3.3). In fact electronics played a minor role as a determinant of growth in interwar Europe and Japan, and even in the USA it was not crucial until the 1950s and the boom in demand for consumer electronics. Nonetheless, early developments in electronics were a significant result of the increasingly closer links between science and technology in innovative activities. A illuminating case is the invention of the transistor, which was the outcome of research carried out by a group of scientists at the Bell Laboratories. The Bell Laboratories of the time were the prototype of research centres in which advances in fundamental science are transformed into viable innovations, and where the link between science and invention becomes highly complex and dialectical (see Rosenberg, 1982, p. 143). After the introduction of R&D laboratories in industrial firms – whose origins can be traced back to the Chemicals technological style described in Section 5.3.2 – a second organisational innovation to have changed the features of research activities was therefore the creation of large research centres in which basic research was undertaken according to the needs of technological research, and in which any new scientific discovery could be rapidly transferred into industrial products. This revolution in the organisation of research activity, which began with the boom following the Second World War, was further boosted by the invention, during the same period, of various synthetic materials (for example nylon, rayon and PVC) and pesticides (such as DDT), which significantly affected the patterns of industrial and agricultural development.

From the financial viewpoint, the breakdown of the international gold standard after 1914, the financial instability that typified the 1920s and the panic of 1929–33 following the Wall Street Crash, gave rise in most countries to the new financial style characterised by the central banks' involvement in the management of public debt and in the control over interest rates and financial intermediation (see Roosa, 1951).[17] Since the main interwar problem was basically one of macroeconomic demand deficiencies, central banks had knock-on effects for the demands for the products of large industrial firms. In this respect, their widespread diffusion can be taken as one of the factors which permitted the development of large plants and the exploitation of scale economies throughout the industrialised world.

In the USA, the power of the Federal Reserve System – originally created to cope with the problems of public debt management in proximity of the First World War – increased greatly. Its ability to alter reserve requirements and the lending power of the Federal Reserve Banks was broadened, and its control over the quantity of money was extended (see Friedman and Schwartz, 1963).[18] Thus, when the price and use of credit also came under its control, the Federal Reserve System acquired a crucial role in the determination of monetary policy, and as a consequence the activities of most financial intermediaries became increasingly subordinated to its regulation and control.

The growing importance of central banks during the period under consideration must be viewed in relation to both the breakdown of the previous institutional arrangement based on the international gold standard and the increasing function of governments in the pursuit of high levels of employment and a stable growth path. During the fourth Kondratieff long wave, central banks were therefore called upon to perform functions directly related to monetary and financial stability. From the 1930s to at least the beginning of the 1970s, these institutions therefore pursued several goals, ranging from control over money growth – mostly through open market operations in which they bought bonds and paid for them with money – to identification of self-regulation mechanisms able to prevent the financial crises, panics and instability caused by the insolvency of some financial institutions. Such control and support functions were widely exercised during the fourth alleged Kondatieff long wave, and over the years were alternately reinforced and slackened by governments in the belief that they could exert some moderating restraint on cyclical fluctuations. In particular, a favourable attitude towards money growth and the acceptance of deficit financing permitted implementation of the Keynesian or, broadly speaking, interventionist policies that helped to foster diffusion of the 'Fordist' technological style prior to the 1970s.

5.3.5 Towards a Fifth Kondratieff Long Wave?

The social turmoil of the late 1960s, followed in the 1970s by the oil shocks and the breakdown of the Bretton Woods fixed exchange rates system, signalled that the 'Fordist' technological style was close to exhaustion. A radical change in the techno-economic subsytem was therefore necessary. In fact, since the second half of the 1970s two latent technological styles have emerged: information technology and biotechnologies.[19] One may now predict that these will become the leading and most pervasive technological styles of the upswing phase of a hypothetical fifth Kondratieff long wave. If some of the chronologies presented in Table 5.1 are extended in a highly deterministic manner, the A period of the fifth Kondratieff should begin by the mid-1990s.

Although it might be contended that both information technology and biotechnologies have spread throughout the industrialised world during the last fifteen years, it would be incorrect to consider them as fully active technological styles. This point is self-evident in the case of biotechnologies, which still represent an infant industry in the international arena, but it is also consistent with the current diffusion of information technology. After their initial adoption on a large scale both at the industrial and the household levels, information technologies are now marking time and the diffusion cluster has not yet reached the magnitude required to revolutionise the techno-economic subsystem. Explanation of the relatively slow diffusion of this new technological style lies mainly in the mismatch between the education system in most countries and the training required to cope with information technology in everyday life. Widespread and radical change in the socio-institutional framework is therefore needed before information technology can become the leading technological style of the alleged fifth Kondratieff upswing. In this connection, the Integrated Services Digital Network (ISDN) represents an important bandwagon that is expected to allow, in the remainder of this century, significant productivity gains in the employment of information resources. ISDN offers:

the prospect of unified routing and efficient transmission of information for virtually any conceivable purpose. For public network companies, ISDN provides a means of market entry into any existing (e.g. cable television) or conceivable (e.g. high definition facsimile transmission) service market. For large businesses, ISDN offers an alternative to increasingly complex and costly piecemeal solutions for data communications problems. For smaller businesses, ISDN promises to preserve meaningful entry opportunities in markets that otherwise might be dominated by larger firms that have invested in private networks. (David and Steinmueller, 1990, p. 44)

The early stages of diffusion of information technology and bio-technologies have been characterised by a shift in the size distribution of firms, which has led to the supremacy of small rather than large firms as vectors of innovation (see Oakey, 1984). This tendency has reversed a century-long process of industrial concentration, at least since the second Kondratieff long wave, and confirms the point made by Schumpeter (see Section 2.3) concerning the declining innovational impetus of large firms and the ability of innovator–entrepreneurs to introduce major innovations. The crisis of the 'Fordist' style determined a shift towards small firms, through a process of restructuring and vertical disintegration that modified in depth the function of larger firms. The main characteristics of small firms and new technology-based firms (NTBFs) belonging to high-tech industries during the years preceding the upswing of the fifth hypothetical Kondratieff long wave can be listed as follows (see also Rothwell, 1985):

– Adaptability to fast-changing market requirements and flexibility, which were previously features only of small firms belonging to traditional industries (see Piore and Sabel, 1984; Santarelli and Sterlacchini, 1990, 1994a).
– Dynamism, entrepreneurial capability and willingness to bear risk as the main characteristics of managements, which are particularly adept at taking advantage of new production opportunities.
– A governance structure of internal transactions based upon a non-hierarchical organisation, and a substantial lack of bureaucracy, which allows the introduction of more efficient internal communication networks.
– Heavy reliance on financing of the venture capital type for start-up or early-stage financing, closely related to the fact that innovation involves these firms in a large financial risk, and short-term loans do

not always allow the undertaking of innovation-oriented R&D strategies reaching maturity only after a long period of time.

Many of these NTBFs have been particularly noteworthy for their autonomous inventive ability. In the USA, in particular, they have also played a crucial role in the early diffusion and further development of new technologies previously developed by large firms. The important function of these NTBFs as vectors of innovation can be demonstrated by using the SPRU data base mentioned in Section 5.2. The evidence presented in Table 5.4 has been obtained by dividing the years between 1945 and 1983 into eight subperiods (1945–9, 1950–4, 1955–9, 1960–4, 1965–9, 1970–4, 1975–9, 1980–3) and then computing the average annual rate of growth from the first subperiod of the number of innovations developed by firms and units belonging to each of the seven size classes considered.[20] The results obtained with this simple procedure clearly demonstrate that between the initial and final subperiod the number of significant innovations developed by smaller firms and units with between one and 199 employees[21] has grown more than that of significant innovations introduced by larger firms and units. Evidently, in the UK, small firms and units have represented the most significant vector of technological change (both in terms of generation and diffusion of innovation) during the relevant period.

However, larger firms do not appear to be entirely losing their function. In particular, it must be remembered that the revolution in personal

Table 5.4 Innovation share by size of firm (i.e. independent business units) and by size of innovating local units (i.e. non-independent business units belonging to firms) to in the UK, 1945–83 (weighted rates of growth)

	Size of firm and of innovating unit						
	1–199	*200–499*	*500–999*	*1000–9999*	*10 000–29 999*	*30 000–99 999*	*100 000 and above*
By size of firm	0.171	0.140	0.109	0.085	0.096	0.104	0.173
By size of unit	0.191	0.160	0.164	0.030	0.047	0.200*	0.055

* This result is due mainly to the very low number of innovations introduced by units belonging to this size class (as compared with the total number of innovations in all size classes) during the initial subperiod 1945–9. When the first subperiod, evidently characterised by the prevalence of smaller units, is not considered in the analysis the weighted index is instead 0.036.

computing and software design initially brought about by NTBFs – who gained ground in terms of total revenues with respect to large established companies – was continued in a second stage by PEPFs (for example IBM) and by relatively new firms that a few years after their start-up had reached a very large size.[22]

The level of investment expenditures by both NTBFs and large companies is determined by the availability of long-term finance and the existence of a large capital market, which in turn is closely dependent upon the more or less efficient functioning of the financial framework and its ability to produce new institutions and instruments. In fact, during the late 1970s and the 1980s a significant process of financial innovation was started in the UK and the USA. As regards financial institutions, this innovative process partly reduced resistance to the initial diffusion of the newly emerging information technology and biotechnologies styles. Moreover the increasing importance assumed by already existing but relatively underexploited institutions such as venture capital firms and pension funds helped to meet the financial requirements of NTBFs and PEPFs in the two industries. The growth of these financial institutions was of course stimulated by a slackened control exerted by central banks, although in the UK and the USA their initial diffusion can be traced back to the 1950s, when thrift organisations, insurance companies, mortgage banks, development banks, venture capital firms, pension funds and investment companies started to grow more rapidly than traditional financial institutions (see Goldsmith, 1971). During the 1950s and the 1960s the growth of these institutions still continued under the control of central banks, in accordance with the principle set out in the 'credit availability' theory (see Roosa, 1951). Conversely, after deregulation got under way in the 1970s, the diffusion process accelerated greatly following a reduction in the control function exerted by central banks and the substantial globalisation of financial markets. Since then, at least in the UK and the USA, venture capital firms and pension funds have greatly increased their functions and if they also become widespread in the other industrialised countries they are likely to represent the most important institutions in the financial style of the next (fifth) alleged Kondratieff upswing.[23]

Since the role played by venture capital firms in the early financing of data processing firms in the 1970s and 1980s will be analysed in more depth in Chapter 7, I conclude with a brief mention of pension funds, which are probably about to become the most important institutional investors on a world scale, apace with the reform of welfare systems already under way in most countries. In effect, through workers' and employers' contributions to pension schemes, the corporate sector is able

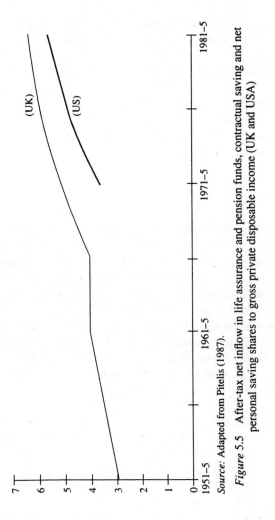

Source: Adapted from Pitelis (1987).

Figure 5.5 After-tax net inflow in life assurance and pension funds, contractual saving and net personal saving shares to gross private disposable income (UK and USA)

to retain the net inflow of pension funds in order to finance its next period's investment (Pitelis, 1987, p. 56). In fact, because of the greater financial activity by pension funds and similar institutions, the proportion of private disposable income retained and invested in the corporate sector has grown significantly in the UK and the USA during the last decades (Figure 5.5). If this process becomes transnational in character, one may predict that – just as venture capital contributed significantly to the early crystallization of the information technology and biotechnology technological styles – the 'pension funds revolution' will help to stimulate the full assimilation of these new technological styles into the techno-economic subsystem. In effect, whereas venture capital firms meet the financial requirements of NTBFs, pension funds channel additional (with respect to the saving of the corporate sector) financial resources towards joint stock companies (PEPFs).

5.4 SUMMARY

This chapter has shown that innovations in the financial system and technological innovations relevant to the production system are likely to be interdependent up to a reasonable degree of association. Simple statistical analysis has shown that basic technological innovations tend to concentrate more during depression periods, whereas financial innovations and events are more typical of boom periods. Although these findings do not provide a thorough demonstration of Schumpeter's view of the introduction of major innovations and of Mensch's 'bunching' theory of basic innovations, they nonetheless support the idea that *basic innovations* do not necessarily occur in boom periods, when it is the *diffusion cluster* that fosters economic growth and development.

On the basis of these results, the second part of the chapter outlined the role of different financial styles in permitting – through the modification of an important aspect of the socio-institutional framework – the diffusion of new technological styles (*diffusion cluster*) able to cause significant change in the techno-economic subsystem. This assumption of inter-dependence goes far beyond the hypothesis of Schumpeter surveyed in Chapter 2, namely that finance and the banking system occupy a pivotal role in technological innovation. It rests, in fact, on the idea that the diffusion of new attitudes and institutions in the financial sphere is a factor that, by helping to restructure the socio-institutional framework, reduces the 'mismatch' between the latter and the new technological style and indirectly stimulates economic expansion.

6 The Structure of the Data Processing Industry in the 1980s and Early 1990s

6.1 DATA PROCESSING DEVICES AS ECONOMICALLY AND TECHNOLOGICALLY STRATEGIC GOODS

The aim of this chapter and of Chapter 7 is to provide empirical confirmation for some of the theoretical assumptions developed in Chapter 4 concerning the financing strategies of innovative firms belonging to high-tech industries. Unfortunately, lack of reliable data on firms' financial structure does not allow for a rigorous econometric investigation. Thus, Chapter 6 first introduces the general features of the data processing (henceforth DP) industry, with especial reference to R&D spending and the role of R&D cooperation agreements. Chapter 7 presents some case studies of successful venture-capital-funded start-ups (NTBFs) in this industry and a simple regression analysis carried out on a sample of large corporations of the PEPF type in which the rates of growth of shareholders' equity and of long-term debt are the explanatory variables, and the rate of growth of R&D expenditures is the dependent variable.

The decision to focus on the DP industry for empirical justification of the theoretical assumptions of Chapter 4 stems from considerations concerning the nature of technology and of financial strategies in this industry. Firstly, from a technological viewpoint the DP industry contributes significantly to the information technology style that was predicted in Section 5.3.5 as being the leading technological style of the next Kondratieff upswing. Secondly, most DP firms relied on financing of the venture capital type to fund their start-up and early stage development. Since venture capital firms were shown in Section 5.3.5 to be among the most important institutions within the newly emerging financial style, in this respect also the decision to study in depth the DP industry appears fully justified.

As far as technological change in the DP industry is concerned, many of the innovations that have revolutionised computing since the early 1970s can be attributed to increasingly higher levels of integration on a single chip. This process of innovation in the core technology has led to the introduction of devices such as the 1-megabyte dynamic random access memories (DRAMs), responsible for the very high rates of price/ performance improvement in computing. Semiconductor technology is therefore the base-level technology that drives innovation in computing. However some of the progress achieved during the last two decades derives from software rather than hardware technology or, in the area of hardware, from innovations not directly linked to semiconductor technology. For this reason, by focusing on the financial strategies of companies belonging to the DP industry in the strict sense – that is, leaving the semiconductor branch aside – it is possible to obtain useful insights into the way innovations can be developed even without significant improvements in the base technology. In effect, most companies in the DP industry have devised innovative financial strategies in order to raise the funds needed to implement 'D'-oriented R&D strategies.

The dramatic technological change in DP technologies over the last two decades can be explained within a traditional demand-pull/technology-push framework. However, although the role of demand factors can be easily defined, that of supply-side factors deserves careful analysis, in particular as regards the non-technological determinants of innovative capability.

The crucial role of demand-side factors were summed up in the words of Intel's cofounder and chairman Gordon Moore on the critical elements responsible for the pattern of innovation in data processing:

> One of the principal things it depends on is the continuing market elasticity we've enjoyed. That when we put 1,000 times as many transistors on a chip, we can find people who want to use a lot more of those chips.... And it's only by the continuing expansion of the markets that we're able to make the investments and keep the technology moving. (as reported in Moad, 1990, p. 26)

Conversely, the role of supply-side factors and the formation of innovative capability cannot be explained solely in terms of the endowment of scientific and technological knowledge. It is true that firms select and implement among a range of technological options only those with which they are most familiar (see Santarelli, 1995), but they may resort to various instruments to foster this process. Identification of the optimal finance structure of the firm as a major factor affecting the potential for innovation therefore directs economic research less deterministically by emphasising a crucial institutional determinant of those activities (in particular R&D) that allow firms to create and develop new technologically advanced and commercially viable products.

This chapter is structured as follows. Section 6.2 surveys the main technological determinants of competition in the DP industry during the late 1980s and early 1990s. Section 6.3.1 presents an analysis of the structure of the DP industry, focusing in particular on concentration ratios. Section 6.3.2 analyses the R&D intensity of this industry in various countries, while Section 6.4 compares the distribution of DP companies in the USA, Europe and Japan. Finally, in Section 6.5 some concluding remarks are made.

6.2 THE COMBINED EFFECTS OF DEMAND-PULL AND TECHNOLOGY-PUSH

The total demand for DP equipment grew along in the 1980s with the availability of more powerful chips, standard operating systems, improved peripherals, new application software and the rapid diffusion of local area networks (LANs). This relationship between technological change and the growth of total shipments of computer devices was close and significant until the end of the decade, when the market entered a phase of slower growth punctuated by the poor performance of some of the most important companies in terms of both total sales and profit rates. Thus in the early 1990s the elasticity of demand for DP equipment to improvements in technology is less marked than in the past,[1] although the economic performance of the industry is still satisfactory.[2]

The DP industry uses semiconductors as the basic component in computing and communications systems. However technological change in the DP field depends both on the availability of more powerful chips and on the ability of firms in the industry to exploit the results of in-house R&D and the creativity of their engineering departments. One striking example of this non-semiconductor dependent path of innovation is provided by compact hard disk drives able to store 1.5 billion bits of information and retrieve any of them in 25 one-thousandths of a second (Kupfer, 1990). Such devices, produced by companies such as Conner Peripherals, have triggered a boom in notebook-size portable computers,[3] which would never have been possible if these companies had relied solely on improvements in semiconductor technology. Accordingly, the present chapter excludes the semiconductor industry from analysis and examines the market and R&D performance of firms belonging to the DP industry. It focuses on the DP industry as a whole, rather than on the computer industry alone, to discuss the increasing importance of the integration of computers, communications equipment, electronic components, systems integration devices and software (see Flamm, 1987).[4] Mainframes, minicomputers and stand-alone personal computers are rapidly approaching maturity, whilst networked workstations, client/server configurations and custom software have boomed during the late 1980s and early 1990s. This process is manifest in the double-digit growth of the systems integration market, which has convinced companies such as IBM and Digital Equipment Corporation (DEC) to devote increased effort to technologies that support distributed computing, multimedia, maintenance of application software and services. Thus, taking for granted the path of technological change in semiconductors and printed circuit boards,[5] it is likely that analysis of the DP industry will highlight the most significant technological, financial and organisational factors in the recent development of information technologies more clearly than if the focus is on the computer industry, in the strict sense, alone.

In the last few years a combined demand-pull/technology-push effect has shifted competition to new areas, thereby emphasising the importance of concepts such as central processing unit performance and price/ performance relationships in data processing. From the technological perspective, the 80486 32-bit chip marketed by Intel has blurred the distinction between PCs and workstations and allowed the introduction of much more powerful high-end microcomputers, which have gained ground in the market for computers at the expense of minis. This tendency has strengthened since the introduction of Pentium, Intel's more recent creation, which has set the new standard in the PC field. On the demand side, buyers have become increasingly price-sensitive and insist on the compatibility of the operating systems and machines to be integrated with their already existing systems.

6.3 SALES, QUANTITIES AND TECHNOLOGICAL CHANGE IN THE DP INDUSTRY DURING THE 1980s

6.3.1 Total Sales and Market Concentration

The worldwide market for DP products measured in total sales increased significantly in the 1980s. However in 1989 the rates of growth of total sales started to slow down in various branches of the industry, with the exception of telecommunications equipment. The data processing industry – and in particular its office equipment branch – is still, in 1994, expected to experience relatively slow growth rates of total demand, at least until the expansion of multimedia brings about a significant recovery. In this connection, the agreement of October 1991 between IBM and Apple (see *The Economist*, 1991) is expected to produce its first tangible results between 1994 and 1996 with the early introduction and subsequent diffusion of a new operating system based on IBM's AIX, adapted to the Macintosh user interface. This is likely to be the first step towards the widening of product variety.[6]

A breakdown of the various branches of the DP industry (Table 6.1) shows that the worldwide market for office equipment grew (in terms of total sales) during the 1980s at an average annual compound rate of 9.3 per cent. This was far below the 19 per cent registered by software, services and maintenance (that is, intangibles) – which have represented the fastest growing market in data processing – and it is also below the average growth rate of the whole industry (10.8 per cent). In this respect, it is worth noting that by 1987 software, services and maintenance had overtaken hardware in terms of total sales. This was due to very different growth rates of total sales for hardware and intangibles: between 1987 and

Table 6.1   Worldwide data processing market (US $bn at 1980 prices)

	EDP equipment	Office equipment	Telecom Equipment	Software services & maintenance	Total
1980	119.9	30.5	79.1	37.5	267.0
1981	130.3	31.3	79.3	44.6	285.5
1982	147.0	36.0	80.8	51.3	315.1
1983	162.4	40.9	85.4	62.3	351.0
1984	187.1	45.0	90.9	77.0	400.0
1985	213.7	48.8	94.6	88.3	445.4
1986	238.2	52.3	99.5	98.6	488.6
1987	270.2	56.8	103.1	111.5	543.6
1988	304.0	62.1	112.2	126.4	604.7
1989	331.8	66.9	118.1	143.1	659.9
1990	357.3	71.6	122.8	158.5	710.0
1991	388.9	76.7	129.5	176.5	771.6
1992	426.4	82.9	137.0	197.8	844.1
1993[1]	464.8	89.8	144.9	221.6	921.1
CGR80–88 (%)[2]	12.3	9.3	4.3	19.0	10.8
CGR88–93 (%)	8.9	7.7	5..3	11.9	8.7

Notes:
1. Expected value.
2. CGR = average annual compound growth rate.
Source: CBEMA, *Industry Marketing Statistics*, The Center for Economic Analysis, Inc.

1993 the growth of hardware was 40.5 per cent, and that of intangibles 98.7 per cent. At least three factors were at work here:

1. A widespread tendency among companies to reduce the total costs of information technology by investing more in systems integration devices than in wholly new computers.
2. Increasing recourse to 'outsourcing', that is, the farming out of critical data processing functions to service providers.[7]
3. A dramatic reduction of investment in information technology by small and medium-sized firms using minicomputers. This process enhanced the average length of life of minis and postponed their replacement.

However, if one inspects the sales figures for mainframes (including supercomputers), minicomputers and microcomputers (either PCs[8] – including desktop, laptop, notebook and handheld computers – or workstations), some differences clearly emerge even in the field of office equipment (see Figures 6.1, 6.2 and 6.3).

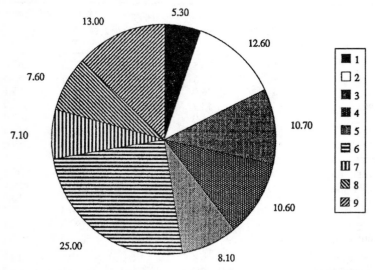

Notes: (1) Other, (2) mainframes, (3) minicomputers, (4) PCs, (5) software, (6) peripherals, (7) data communications, (8) services, (9) maintenance.

Source: Datamation (1988).

Figure 6.1   Shares of DP market by product line (1987, percentages)

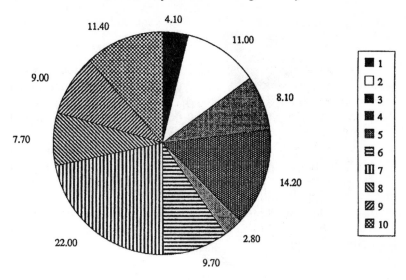

Notes: (1) Other, (2) mainframes, (3) minicomputers, (4) PCs, (5) work-stations, (6) software, (7) peripherals, (8) data communications, (9) services, (10) maintenance.

Source: Datamation (1990).

Figure 6.2   Shares of DP market by product line (1989, percentages)

For example, PCs significantly increased their market share between 1987 and 1991, whereas problems arose in the market for mainframes (whose share diminished from 12.6 per cent in 1987 to 9.5 per cent in 1991), and, in particular, in that for minis (from 10.7 per cent to 7.6 per cent). This tendency toward downsizing becomes even more significant if one considers the fact that workstations, which did not exist in 1987 and were initially introduced by Sun Microsystems (still the leader in this field), represented 2.8 per cent of the total DP market in 1989 and 4.7 per cent in 1991. PCs and workstations are being adopted by user companies that formerly relied on minicomputer platforms to build their DP infrastructures. Buyers of minis seem to hesitate over new purchases even more than those of micros and mainframes, essentially because of increasing competition from high-performance workstations. However it generally appears that the ability of users to absorb newer generations of technology is declining: in the field of midranges (minicomputers), the introduction of reduced instruction set computing (RISC) and of graphic

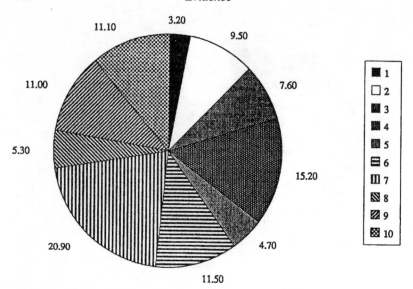

Notes: (1) Other, (2) mainframes, (3) minicomputers, (4) PCs, (5) work-stations, (6) software, (7) peripherals, (8) data communications, (9) services, (10) maintenance.

Source: Datamation (1992).

Figure 6.3   Shares of DP market by product line (1991, percentages)

user interfaces (GUIs) has increased competition between companies such as IBM, DEC and Hewlett-Packard without fostering the growth of this market.

But this is also true of the market for PCs, where sales of Intel 80486-based machines have initially outstripped even those of the most powerful Pentium offerings, and where the first half of the 1990s has seen a slowdown in the rates of growth of total sales.

As regards mainframes – a branch in which IBM still holds the dominant position, with over 70 per cent of the US market in 1990 – technological competition has principally centred on supercomputers,[9] which represent the strategically most important class of computers, although in terms of total shipments they account for only a small portion of the market. Traditional mainframes, conversely, are undergoing a dematurity process, which in the most important producers' expectations should make its effects felt by 1996, with the possible resurgence of these products in the market. The main losers in the race for control of the data processing market are apparently minicomputers (a field initially dominated by IBM and DEC), therefore, since they suffer from a low price/performance ratio compared with high-end microcomputers. The field of midrange computers is likely to be taken over by the most powerful PCs and workstations, which will continue to gain momentum in the foreseeable future: it is predicted that by the end of 1995 in the USA more than 85 per cent of the PCs within organisations will be networked.

A significant increase in market share has occurred in the case of software, which accounted for 8.1 per cent of the DP worldwide market in 1987 and 11.5 per cent in 1991. This growth is even more important if one considers that it has been led by hardware producers such as IBM, Fujitsu and NEC, which are differentiating their production in order to maintain leadership in the DP market.[10] The same applies to services. Here the increase from 7.6 per cent of the total market in 1987 to 11 per cent in 1991 is partly explained by the rapid growth in the field of services of traditional hardware suppliers such as IBM, DEC and Fujitsu, which are now competing for market leadership with Electronic Data Systems (the world leader in systems integration and outsourcing), Andersen Consulting, TRW, Computer Sciences Co. and others.

Generally speaking, the DP industry's main problems stem from the progressive maturing of its products in terms of performance variety and from a lack of confidence among users, who demand high standards and the certainty that their software has maximum portability. As a matter of fact, recent developments in computer technology have not affected the performance of data processing devices for the most common applications. The important advances of recent years in networking and systems integration have benefited the small portion of users represented by large companies and the government, whereas smaller companies can still operate their previous, poorly integrated data processing systems.

The race for compatibility, the rise of new leading companies such as Compaq and Sun,[11] and the recent crisis, have not led to significant concentration within the industry. In particular, the *concentration ratios* presented in Table 6.2 demonstrate that between 1984 and 1991 the share of total revenues by the first four (C_4), the first eight (C_8) and the first twenty (C_{20}) companies in the data processing industry initially declined (until 1988) and then started to grow slightly in the next two or three years.[12] Conversely, the share of the first company (IBM) declined dramatically, from 33.5 per cent in 1984 to 21.7 per cent in 1991.

However, the concentration ratio (C_k) is a poor proxy for the dynamics of market structure, since it ignores the differences in terms of market shares among the K largest companies in a given market. This limitation

Table 6.2   Concentration ratios (*C*) and Linda index (*L*) for the world data processing market as measured by the Datamation ranking (1984–91)

	C_1	C_4	L_4	C_8	L_8	C_{20}
1984	33.5	44.5	**1.68**	55.1	**0.66**	72.4
1985	30.8	42.0	**1.59**	52.5	**0.74**	70.5
1986	28.0	37.1	**1.19**	53.1	**0.54**	71.0
1987	24.2	35.5	**1.27**	49.6	**0.47**	67.7
1988	22.7	36.7	**0.88**	48.9	**0.44**	68.1
1989	23.7	37.7	**0.89**	50.3	**0.44**	69.3
1990	24.3	37.9	**0.92**	50.9	**0.41**	69.5
1991	21.7	38.5	**0.90**	51.4	**0.42**	67.2

Source: Elaboration on *Datamation* (various issues).

may be overcome by using the *Linda index*, which allows measurement of the degree of inequality in the values of a variable included in a subsample of *K* variables. The Linda index (*L*) is represented as

$$L = \frac{1}{K(K-1)} \sum_{1}^{k-1} Q_i$$

with

$$Q_i = \frac{A_i}{i} \bigg/ \frac{A_K - A_i}{K - i}$$

where *K* denotes the sub-sample of large firms in the market, and Q_i is the ratio of the cumulative percentage of market share of the first *i* firms to the average market share of the remaining *K* − 1 firms. When used for comparing market shares, a decrease in the Linda index between a given year and the next one denotes a decrease in the degree of inequality in the market shares of firms included in the sub-sample *K*.

In Table 6.2 the Linda index is computed for each year from 1984 to 1991 in relation to the first four (L_4) and the first eight (L_8) firms included among the *Datamation 100*. The L_4 index shows a significant decrease in the degree of inequality for the first three years: in particular, the most dramatic drop occurs in 1986 and can be attributed to the merger between Sperry and Burroughs, which in that year gave rise to Unisys (ranked second). Conversely, the index growth in 1987 is due to the low level of DP revenues accounted for by Hewlett-Packard, which ranked fourth in that year. L_4 remains substantially unchanged between 1988 and 1991, with minor changes from year to year. During this period of stability, the slight fall of the Linda index in 1991 is instead due to the poorer performance of IBM with respect to the previous year and to the simultaneous growth of Fujitsu and NEC, ranked second and third respectively.

These findings partly change the picture with respect to the concentration ratios, since they reveal a very significant reduction of inequalities in terms of market power among the four largest companies in the DP market. Even more interesting is index L_8, which becomes effectively stable after 1987, following the significant drop in 1986 brought about by the creation of Unisys. The reduction in L_8 over the last two years considered by Table 6.2 can instead be attributed to the acquisition of Nixdorf by Siemens, which in 1990 gave rise to the largest DP company in Europe.

The Linda index can also be usefully employed to identify the oligopoly threshold. The procedure here is to compute the Linda index for the first two ($K = 2$), the first three ($K = 3$), and the first n ($K = n$) firms until a minimum value of L is obtained; that is, until the L index for $K + n$ firms is higher than the analogous index for $K + (n - 1)$ firms. In this case the 'oligopoly area' includes $K + (n - 1)$ firms. Unfortunately, in the case of the DP firms included among the *Datamation 100*, the application of this procedure does not yield satisfactory results, since the distribution of market shares accounted for by each firm decreases monotonically. Thus these data cannot be used to identify any 'oligopoly area' in the DP market.

6.3.2   The Dynamics of R&D Expenditures: International Comparisons

In the 1980s and early 1990s technological change in data processing followed a downsizing pattern. The main non-semiconductor-related technological advances occurred in the fields of compatibility and software portability with a consequent dramatic reduction in the price per MIPS (millions of instruction per second), which for workstations is now below $1000. These changes – which are at least partially explained by the process of organisational innovation still ongoing within the industry – have been made possible also – but not only – by an increase in the total funds devoted to R&D. In particular, inspection of total expenditures on R&D (Table 6.3) in the office machinery and computers industry for each of the OECD countries reveals that between 1981 and 1988 R&D

Table 6.3 Office machinery and computers: total R&D expenditures (in millions of constant US $)

	1981	1982	1983	1984	1985	1986	1987	1988
Australia	33.9[3]	–	–	75.8[3]	–	194.9[3]	–	254.6[3]
Austria	41.1	–	–	28.4	–	–	–	–
Belgium	0.2	–	0.3[4]	–	0.4[4]	0.4[4]	0.5[5]	7.0[5]
Canada	83.0	112.8	121.5	143.8	157.1[1]	190.6	205.7	216.8
Denmark	12.2[2]	–	13.7[2]	–	13.4[2]	–	19.8[3]	–
Finland	10.6	–	11.0	–	–	–	–	–
France	326.1	340.6	323.7	383.1	427.0	443.3	431.2	418.9
Germany	–	–	–	–	–	–	–[5]	–
Greece	–	–	–	–	–	–	–	–
Iceland	0	–	0.045	–	0.053	–	0.07	–
Ireland	–[5]	–[5]	–[5]	–[5]	8.47	15.96	11.27	6.7
Italy	184.7	185.1	192.7	243.7	290.4	288.7	303.1	303.7
Japan	–[5]	–[5]	–[5]	–[5]	–[5]	–[5]	–[5]	–[5]
Luxembourg	–	–	–	–	–	–	–	–
Netherlands	–[5]	–[5]	–[5]	–[5]	–[5]	–[5]	–[5]	–[5]
New Zealand	–	–	–	–	–	–	0.2	–
Norway	11.0	10.8	14.8	22.0	26.2	–	37.2	–
Portugal	–	–	–	–	–	–	–	–[5]
Spain	12.5	13.9	4.1	17.3	58.7	53.9	80.7	–
Sweden	–	–	54.9	–	125.8	–	141.6	–
Switzerland	3.7	–	–	–	–	–	–	–
Turkey	–	–	–	–	–	–	–	–
UK	385.6	–	502.7	–	630.0	624.7	624.5	681.7
USA	–[5]	–[5]	–[5]	–[5]	–[5]	–[5]	–[5]	–[5]
Yugoslavia	–	–	–	–	–	–	–	–

Notes:
1. Break in series with previous year for which data is available.
2. Secretariat estimate or projection based on national sources.
3. National estimate or projection adjusted if necessary by the secretariat to meet OECD norms.
4. Underestimated or based on underestimated data.
5. Included elsewhere (that is, in other industries).
– = data not available.
Source: OECD (1991).

expenditures in this industry grew (at constant prices) by 651 per cent in Australia (!), 76.8 per cent in the UK, 64.4 per cent in Italy and 28.46 per cent in France.

Although the level of R&D expenditures is a good proxy for an industry's commitment to innovative activities, more reliable information can be obtained by using an index of R&D intensity, such as the ratio of R&D expenditures to total sales (see Table 6.4).

Table 6.4 R&D intensity in the DP industry

	1981 %	1982 %	1983 %	1984 %	1985 %	1986 %	1987 %	1988 %
Australia	1.81	–	–	5.45	–	12.05	–	15.84
Canada	5.88	8.45	9.12	8.20	8.36	10.80	10.71	8.86
Denmark	14.19	–	12.53	–	10.58	–	18.88	–
Finland	8.79	–	7.26	–	–	–	–	–
France	11.89	12.41	10.20	10.09	10.01	10.76	–	–
Germany	–	–	–	–	–	–	–	–
Greece	–	–	–	–	–	0.10	–	0.19
Italy	5.02	4.65	5.40	5.07	4.47	4.99	4.94	–
Japan	24.90*	–	23.40*	20.10*	21.30*	25.20*	25.20*	–
New Zealand	–	–	–	–	–	–	0.41	–
Norway	8.15	6.87	8.24	8,75	9.16	–	12.91	–
Spain	2.89	2.73	0.96	1.75	4.67	5.53	8.78	–
Sweden	–	–	10.33	–	15.97	–	17.61	–
UK	11.97	–	11.21	–	8.32	8.70	7.10	5.69
USA	27.30*	25.50*	24.90*	22.20*	23.20*	25.30*	26.40*	24.60*

Notes:
– = data not available.
* Total sales data for the whole electrical group.

Source: Elaboration on OECD (1991).

Note that it was not possible to provide in Table 6.3 the R&D figures for, among others, Germany, Japan and the USA, since these data are provided by OECD at a more aggregated level. In order to overcome this problem Table 6.4 uses production data for the whole electrical group of such countries. Although this procedure does not allow for comparisons among countries, it nonetheless enables a raw figure for the performance of each country over time to be calculated.

The results are extremely interesting. First, between 1981 and 1988 the R&D intensity decreased for France, Italy, the USA (with the caveat of the aggregate nature of data employed for this country) and, more markedly, the UK. Second, R&D intensity in Japan (even if the figure refers to the whole electrical group) grew during the relevant period only by 1 per cent – much less, therefore, than one might expect from Japan's economic performance in the same years. Third, the most significant growth of the R&D to sales ratio occurred in countries without a well-established data processing industry, such as Australia,[13] Denmark and Spain. The only countries for which growth of the R&D intensity presumably denotes real progress in terms of technological capability are thus Canada, Norway and Sweden.

The figures in Table 6.4 demonstrate that, at least at the industry level, R&D expenditures can hardly be taken to be the sole determinant of the process of product innovation that characterised the data processing arena in the 1980s. In this respect it is likely that many new products stemmed from greater innovative activity in firms' design and production departments rather than from *formal* R&D activities in the strict sense. Such *informal* R&D – which is not accounted for by the official R&D surveys – presumably positively affected the internal levels of innovative activity in many firms belonging to the DP industry, even when they were not engaged in systematic research activity.

Nonetheless the results of Table 6.4 also confirm the finding of Chapter 4 concerning cross-region and even cross-country differences in the rates of diffusion of new technological styles. The high growth rates of R&D intensity in countries such as Australia, Canada, Denmark, Norway, Spain and Sweden suggest that the information technology (data processing) style initiated in the USA is now spreading throughout all the industrialised countries, which are apparently converging to a common standard.

6.4   THE GEOGRAPHICAL DISTRIBUTION OF DP FIRMS: USA, EUROPE, JAPAN AND THE PACIFIC RIM

Technology-driven industries usually tend to develop around important academic and scientific centres: this was the case with the early development of the DP industry in Northern California (close to Stanford University and the University of California at Berkeley) and Eastern Massachusetts (close to MIT and Harvard University). However, as technology advances and the worldwide market for the industry constantly expands, in many cases geographical epicentres become less important and new producers are likely to locate in different regions and different countries. Nevertheless, this process has only partly affected the DP industry, with some non-American new entrants or rapidly growing companies gaining a significant share of the worldwide market.

Better understanding of the main changes in the international structure of the DP industry can be gained from Table 6.5, which presents the distribution by geographical area of the first 100 firms included in the *Datamation* ranking between 1984 and 1993.[14] These data show very clearly that the slight fall in the number of US companies[15] in the *Datamation 100* is due mainly to the entry by Japanese and non-Japanese Asian producers of DP devices.[16] Since Japanese strength in this industry can be taken for granted, the performance of non-Japanese Asian companies deserves more careful analysis. The first of these companies to be included in the *Datamation 100* ranking in 1986 was Samsung Electronics, a South Korean manufacturer of integrated electronics. However this company soon encountered severe problems in its telecommunications and computer divisions and dropped out of the ranking after only one year. Thus, between 1987 and 1993 it was the Taiwanese DP industry that experienced high growth rates[17] and eventually had three companies included among the *Datamation 100*: Acer, Tatung and the Mitac Group. The DP industry in the European

Table 6.5   Geographical distribution of DP companies included among the *Datamation 100* (1984–93)

	USA	Canada	EU	Rest of Europe	Japan	Rest of Asia & Pacific Rim
1984	74	–	16	3	7	–
1985	66	2	17	3	12	–
1986	66	–	15	3	15	1
1987	59	1	18	4	17	1
1988	60	1	19	3	16	1
1989	60	2	18	2	16	2
1990	61	2	17	1	16	3
1991	64	2	13	–	18	3
1992	65	2	13	–	17	3
1993	64	2	13	–	17	4

Source: Elaboration on *Datamation* (various issues).

Union has tended to consolidate, with leading companies such as Nixdorf/Siemens, Olivetti and the Groupe Bull coming under increasing competition from American and Japanese manufacturers.

Since the late 1980s EU companies have apparently been unable to keep pace with technological innovation coming from the USA: according to industry analysts, these companies have been relatively slow to implement new technologies (see Appleton, 1991). This loss of technological and market power is confirmed by the drop from the 19 EU companies included in the *Datamation 100* in 1988 to only 13 in 1993; and it also characterised the other European countries, which experienced a drop from four companies in 1987 to none after 1991. Most of the European 'losers' suffered a fall in total revenues brought about mainly by bad performance on the domestic market. For example Amstrad, the Essex (UK) based manufacturer of desk-top PCs, was overtaken on the domestic market by Compaq and Dell, whereas Nokia Data AB, the information technology division of the Finnish conglomerate Nokia, suffered from increasing competition by European, US and Japanese vendors on the Scandinavian market.

6.5 SUMMARY

The 1980s saw the growing importance of the DP industry as a strategic sector, one in which the rate of technological advance outstripped that achieved in other advanced industries. Technological change was mainly initiated by US companies, even though increasing competitive pressure was applied by Japanese, South Korean and Taiwanese companies as the new style spread throughout the world. In the struggle to maintain or achieve leadership of this strategic industry, US companies still dominated the market, although slower rates of growth in R&D expenditures might be a medium- to long-term threat to their technological superiority. Nonetheless the DP industry as a whole underwent a maturation process similar to that experienced on several occasions by the automobile industry in its long history: market saturation and an overabundance of suppliers apparently signal industrial aging in a market where too many companies try to sell new machines to customers who already have one.

7 The Financing of Innovative Activities in the Data Processing Industry

7.1 CREDIT CREATION AND TECHNOLOGICAL CHANGE

As shown in Chapter 6, the initially rapid growth of the DP industry was mainly due to a continuous process of development and diffusion of new products, through a 'D'-oriented R&D activity in most cases set in motion by new firms and reinforced by previously existing ones. In this chapter the relationship between sources of financing and technological activities in the DP industry is outlined in order to provide empirical evidence for the theoretical analysis conducted in Chapter 4. Thus Section 7.2 presents an international comparison of the size and features of the venture capital industry, which, besides its significant function in the early development of the microcomputer industry, has always represented a major source of finance for the whole of the DP industry in the USA. Section 7.3 examines the impact of venture capital on the development of the DP industry in the USA by presenting two case studies concerned, respectively, with successful and unsuccessful venture-capital-funded start-ups. Section 7.4 presents a formal test of the hypothesis introduced in Chapter 4, namely that equity financing is among the preferred sources of financing for large established DP companies investing heavily in R&D. Section 7.5 investigates the wave of takeovers and cooperation agreements that interested the industry during the 1980s and which, in some cases, enabled companies to overcome liquidity problems and to share risk in the undertaking of innovative projects. Finally, Section 7.6 summarises the main results obtained in this chapter.

7.2 VENTURE CAPITAL IN EARLY-STAGE FINANCING OF DP FIRMS: INTERNATIONAL COMPARISONS

Venture capital – the very general features of which were sketched in Section 4.2 – is probably the most expensive form of capital, with discount rates of 50 per cent per year or even more, and it is usually set aside over a long period.

The venture capital industry provides financial and intangible resources to entrepreneurs involved in activities with high growth potential in order to achieve high rates of return on invested funds when the initial public offering (henceforth IPO) is made and the company goes public. There is a sharp distinction between the USA, the UK, Continental Europe and Japan in the recourse to venture capital as a source of financing by DP companies, either for research and product development or for initial marketing purposes. The role of the venture capital industry in the financing of high-tech start-ups – as anticipated in Section 5.3.5 – has been extremely significant in the USA and the UK, whereas this financial institution has operated on a much smaller scale in the other European countries and in Japan.

Table 7.1 provides a worldwide survey of the total capital disbursed by venture capital firms in years 1987 and 1988.

These data confirm the overwhelming importance of the UK and the US venture capital industry which, during the relevant years, provided more than 76 per cent of the total venture capital financing disbursed on a worldwide scale. The result is consistent with the high number of high-tech start-ups that enter the UK and the US market at any time, and which in some cases, mostly in the USA, become successful companies influencing the path of technological change in their industry. Of particular interest is the case of the UK: according to the data presented in Table 7.1 this country is the second single national market for venture capital after the USA, whereas it is the undisputed leader as regards venture capital intensity (that is, in terms of the ratio between venture capital disbursement and resident population). This feature of the UK financial market is closely connected to the process of new firm formation which, during the 1980s, affected high-tech sectors in the country and engendered significant locational clusters such as the M4 Corridor and East Anglia. In this respect, one cannot avoid pondering on why the UK, which has grown so fast in terms of venture capital, has achieved little growth of DP and other information technology-related industries. A plausible explanation is a failure in passing from NTBF to PEPF status, with most innovative start-ups exiting the market before they achieve the minimum efficient scale.

Conversely, venture capital is relatively underdeveloped in the remaining European countries, with the partial exception of France and The Netherlands. For most European countries, the lack of a developed venture capital market may be a plausible explanation for the overwhelming importance of large and established companies with respect to small start-ups in advanced industries. The Italian case provides support for this

Table 7.1 Global VC assets and flows. Pool of capital (1987–8)

	Pool of capital 1987 ($ million)	% Region	% World	Pool of capital 1988 ($ million)	% Region	% World
North America						
USA	29 020	93.2	57.9	31 140	92.3	53.9
Canada	2 123	6.8	4.2	2 585	7.7	4.5
Subtotal North America	31 143	100	62.1	33 725	100	58.4
Europe						
UK	9 257	56.6	18.5	11 531	56.1	20.0
France	2 473	15.1	4.9	3 247	15.8	5.6
Netherlands	1 180	7.2	2.4	1 472	7.2	2.5
Italy	752	4.6	1.5	1 033	5.0	1.8
Germany	590	3.6	1.2	763	3.7	1.3
Belgium	613	3.7	1.2	653	3.2	1.1
Spain	409	2.5	0.8	516	2.5	0.9
Rest of Europe	1 084	6.6	2.2	1 326	6.5	2.3
Subtotal Europe	16 358	100	32.6	20 541	100	35.6
Asia & Pacific Rim						
Australia	649	27.3	1.3	774	24.7	1.3
Japan	1 680	70.6	3.4	2 224	71.1	3.9
Rest of Asia and Pacific Rim	50	2.1	0.1	130	4.2	0.2
Subtotal Asia & Pacific Rim	2 379	100	14.5	3 128	100	15.2
Rest of World	250	100	0.5	350	100	0.6
Total World	50 130	100	100	57 744	100	100

Source: Elaboration on Sahlman (1990).

hypothesis. In effect, despite the significant presence of small and newly established firms throughout the Italian economy, during the second half of the 1980s the index of new firm formation in the Italian DP industry was far below the average index of the manufacturing sector as a whole (see Santarelli and Sterlacchini, 1994a; Piergiovanni and Santarelli, 1995).

Moving to Japan and the Pacific Rim, the situation does not change significantly compared with Europe. However, in the case of Japan the peculiarities of its finance process and the presence of large groups (*keiretsu*), made up of dozens of industrial and financial companies (see Peck, 1988), mean that the existence of a structured venture capital industry is of secondary importance (see Section 7.2.3 below). Japanese industrial enterprises can obtain financial backing much more easily than their counterparts in other countries because they can rely on financial institutions belonging to the *keiretsu*. Thus, albeit in a fashion that differs significantly from that predominant in the USA, in Japan, too, risk capital is readily available for the exploration of new technological horizons.

Table 7.2 presents, for 1987 and 1988, the amount of *new* capital invested by professional venture capitalists in the worldwide market. The picture does not differ significantly from that yielded by the stock data presented in Table 7.1. In particular, the position of the UK appears even stronger, with this country far exceeding the USA in terms of the ratio between venture capital disbursement and resident population.

Table 7.2   Global VC assets and flows. New capital (1987–8)

	New capital 1987 ($ million)	% Region	% World	New capital 1988 ($ million)	% Region	% World
North America						
USA	4 200	85.1	46.4	2 900	86.9	34.6
Canada	738	14.9	8.1	438	13.1	5.2
Subtotal North America	4 938	100	54.5	3 338	100	39.9
Europe						
UK	1 889	53.2	20.8	2 273	54.4	27.2
France	599	16.9	6.6	773	18.5	9.2
Netherlands	113	3.2	1.2	292	7.0	3.5
Italy	134	3.8	1.5	280	6.7	3.3
Germany	361	10.2	4.0	173	4.1	2.1
Belgium	65	1.8	0.7	40	1.0	0.5
Spain	64	1.8	0.7	107	2.6	1.3
Rest of Europe	324	9.1	3.6	242	5.8	2.9
Subtotal Europe	3 549	100	39.2	4 180	100	49.9
Asia & Pacific Rim						
Australia	60	11.7	0.7	125	16.6	1.5
Japan	419	81.5	4.6	544	72.1	6.5
Rest of Asia and Pacific Rim	35	6.8	0.4	85	11.3	1.0
Subtotal Asia & Pacific Rim	514	100	14.5	754	100	18.0
Rest of World	60	100	0.7	100	100	1.2
Total World	9 061	100	100	8 372	100	100

Source: Elaboration on Sahlman (1990).

Conversely, a certain relative weakness emerges in the US venture capital industry, which in 1988 was less involved than in the previous year in the financing of new initiatives. The figures for the other countries were substantially unchanged, although in 1988 new capital disbursements grew noticeably in France and Japan.

We may now concentrate on the three most important markets for the DP industry – the USA, Europe and Japan – in order to outline similarities and differences among them as far as the finance process is concerned. We also take a closer look at the relationship between high-tech start-ups and the availability of financial instruments of the venture capital type.

7.2.1 USA

Capital costs are higher in the USA than in Europe and Japan and have usually been included among the determinants of the short time horizon upon which American entrepreneurs and managers focus when choosing investment projects. According to this interpretation, the managers of American companies are forced to reject projects with long-term payoffs and prefer those that yield quicker payoffs. However, as pointed out by Sahlman (1990), this interpretation contrasts sharply with the success of the American venture capital industry in creating and launching companies that, by investing in long-term projects (between five and seven years on average), have achieved a leading position in international markets. In effect the DP industry in the USA has traditionally exploited this source of capital, at least as far back as the nearly simultaneous launching, in 1957, of Control Data Co. and Digital Equipment Co. (see Katz and Phillips, 1981).[1]

In the USA a venture capital market did not exist in its present form before the Second World War,[2] and its 'invention' is usually traced back to the American Research and Development Corporation (henceforth ARD), a publicly traded independent venture firm created by General Georges Doriot, a professor at Harvard University.[3] Doriot devoted great effort to convincing ARD's public investors (in particular fiduciary institutions) of the importance of nurturing young companies through their early growth stages. However the long-term nature of this new and risky type of investment was not fully accepted by investors, and when the value of ARD's stock companies failed to climb in the short term the price of ARD's stock declined. As a consequence of ARD's difficulties, other venture capital firms were discouraged from going public, and the venture capital market began to develop significantly only when the Small Business Investment Act of 1958 provided for the licensing of small business investment companies (henceforth SBICs). These offered a combination of private equity capital with government-subsidised debt for subsequent rounds of investment in small firms. SBICs were able to obtain up to $300 000 from the government for each $150 000 in private capital and benefited from significant tax advantages. In the venture capital market, experience with publicly traded companies of this type was satisfactory at least until 1962, when investors became less enthusiastic about small electronics and aerospace firms, and the value of SBICs portfolios started to decline. Thus a process of consolidation, punctuated by a number of capital devaluations and liquidations, characterised SBIC industries, and in 1968 for the first time more capital was disbursed by corporate subsidiaries than by publicly traded companies.[4]

The alteration of the industry structure brought by the crisis of publicly traded companies induced a dramatic shift in the venture capital market: between 1969 and the second half of the 1970s the level of new capital committed to and disbursed by this industry steadily declined (see Figure 7.1). During these difficult years syndication became very common, a practice whereby a venture capital firm acted as lead investor and enlisted other firms to coinvest in order to lessen the risk for each investing firm and to attract investors with valuable expertise.

The slump of the market was reversed between 1977 and 1983, when the venture capital industry not only recovered from its poor performance

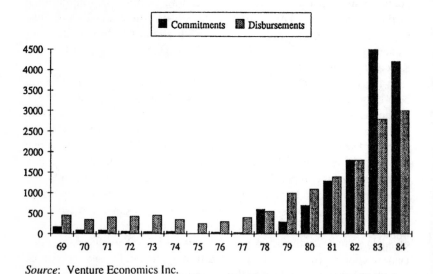

Source: Venture Economics Inc.

Figure 7.1   Total US venture capital industry activity (1969–84)

of previous years, but indeed skyrocketed. According to industry analysts, the 'watershed event' (as defined by Soussou, 1985) in 1977 was the formation of Apple Computer (analysed in detail in Section 7.3.1 below), an event that led to the long-term commitment of venture capital firms to new electronics and computer companies exploiting the potential of the microprocessor chips introduced by Intel at the beginning of the 1970s. New opportunities in the market for microcomputers had knock-on effects in various related industries, including software production (see the case of Lotus Development Co. briefly outlined in Section 4.2) and peripherals (see Section 7.3.3 below).

The technological revolution in the field of information technology was accompanied by a revolution in the financial sphere – with the former calling upon the latter – both of which events occurred initially on the west coast of the USA (see Table 7.3). The second half of the 1980s, after a freeze in 1983, saw a new explosion in the level of activity in the US venture capital market. The total capital pool increased nearly eightfold from 1980 to 1989, and the number of venture capital firms active in the market doubled in the same period (see Table 7.4).

In the early 1980s this further development of venture capital increasingly benefited innovative start-ups, not only in the 'traditional' field of microcomputers but also in the semiconductor and telecommunications industries. Other sectors, however, such as energy, medical products (including biotechnology) and factory automation received a decreasing portion of total funds (see Table 7.5).

By 1981 private independent venture capital firms had largely replaced corporate subsidiaries to become the largest segment in this sector, with 44 per cent of invested capital. The decline of venture capital activities in

Table 7.3   US venture capital disbursements by geographical area (1970–83)

	1970–9 (%)	1981 (%)	1982 (%)	1983 (%)
West coast	30	42	48	52
Northeast	33	25	26	24
Southwest	12	16	13	10
Midwest	12	8	8	7
Southeast	10	5	5	7
Other	–	4	–	–

Note: Figures represent percentages of total disbursement in each year/decade.
Source: Elaboration on Rind, 1985.

Table 7.4   Aggregate US venture capital industry statistics (1980–9)

	(1)	(2)	(3)	(4)
1980	4 500	NA	NA	700
1981	5 800	NA	NA	1 300
1982	7 600	331	1 031	1 800
1983	12 100	448	1 494	4 500
1984	16 300	509	1 760	4 200
1985	19 600	532	1 899	3 300
1986	24 100	587	2 187	4 500
1987	29 000	627	2 378	4 900
1988	31 100	658	2 474	2 900
1989	33 360	674	2 558	3 107

(1) Total venture capital pool ($M).
(2) Number of venture capital firms.
(3) Number of industry professionals.
(4) Net new commitments to the venture capital industry ($M).
Source: Elaboration on Sahlman (1990).

Table 7.5   US venture capital investments by industry (1980–3)

	1980 (%)	1981 (%)	1982 (%)	1983 (%)
Computers	26	30	42	46
Electronics	10	12	14	10
Telecommunications	11	11	10	13
Genetic engineering	8	7	3	3
Medical/health	9	6	6	9
Energy	20	10	6	3
Industrial automation	3	4	3	2
Industrial products	2	5	4	2
Consumer related	2	5	5	7
Other	8	10	7	5

Note: Figures represent percentages of total disbursements in each year.
Source: Rind (1985).

corporations coincided with a growing tendency by new entrepreneurs to exploit the growth potential of their recently formed companies, rather than try to sell new technology or new products to the corporation that had provided them with early-stage financing. Thus, during the 1980s the main sources of funds for venture capital firms were pension funds and individuals, with corporations, insurance companies, endowments and foreigners playing a secondary role. In particular, the amount of capital controlled by the pension funds sector increased sharply after the US government decided in 1979 to liberalise the rules regulating investment decisions by such institutional investors.

7.2.2 Europe

Table 7.6 presents the sectoral distribution of venture capital funds in European countries in the second half of the 1980s. One notes immediately that the amount of capital invested by venture capitalists in Europe grew markedly between 1986 and 1990. Surprisingly, the DP industry is not among the principal recipients of venture capital funds (and this is also true for the UK); indeed the amount of funds obtained by this industry decreased substantially between 1986 (11.7 per cent) and 1990 (7.13 per cent).

Thus in 1990 the DP industry ranked only sixth among the leading recipients of venture capital in Europe, preceded by consumer-related productions, miscellaneous manufacturing, industrial products and services, miscellaneous services, and transportation. However stagnation in venture capital disbursements typified all the high-tech sectors in Europe. These benefited from venture capital to a much lesser extent than one would have expected on the basis of the US experience. As suggested in Section 7.2, this phenomenon was probably one of the causes of the much slower entry process into advanced industries in most European (with the important exception of the UK) countries compared with the USA. The lack of a significant number of innovative start-ups in the DP industry presumably stems from a financial style characterised by large banks being reluctant to fund the formation of small firms exploring the newly emerging technological style. Accordingly, the presence of a well-developed venture capital market may explain why innovative start-ups in the DP industry are more frequent in the UK than in other European countries.

7.2.3 Japan

The development of a venture capital market in Japan was for many years closely connected to the diversification strategies pursued by large financial institutions in that country. Thus Japanese venture capital firms were usually reluctant to take any risk in helping the early start-up of innovative companies, which were all medium-sized firms, growing sluggishly and with little chance of going out of business (see DEFTA, 1990). This picture only began to change in the early 1980s, when a

Table 7.6 Venture capital in Europe by sector of use, 1986–90 (in thousands of current US$)

Industrial sector	1986 (Amount)	(%)	1987 (Amount)	(%)	1988 (Amount)	(%)	1989 (Amount)	(%)	1990 (Amount)	(%)	Total (Amount)	(%)
Communications	66 871	3.54	73 784	2.25	74 880	1.84	142 695	3.04	140 720	2.68	498 950	2.60
Computer related*	220 905	11.7	249 474	7.61	368 479	9.05	397 964	8.48	374 494	7.13	1 611 316	8.40
Other electronics related	146 420	7.76	269 780	8.23	193 119	4.74	133 019	2.83	193 626	3.69	935 965	4.88
Biotechnology	32 212	1.71	66 039	2.02	83 091	2.04	158 857	3.38	121 052	2.31	461 251	2.41
Medical/health related	66 230	3.51	128 460	3.92	123 911	3.04	126 967	2.70	204 064	3.89	649 632	3.39
Energy	44 717	2.37	48 857	1.49	18 900	0.46	52 155	1.11	32 147	0.61	196 776	1.03
Consumer related	359 836	19.07	648 349	19.79	1 011 150	24.83	1 461 976	31.15	1 273 278	24.26	4 754 589	24.79
Industrial products & services	305 188	16.17	552 321	16.86	255 833	6.28	257 446	5.48	435 675	8.30	1 806 463	9.42
Chemicals & materials	63 966	3.39	85 228	2.60	140 610	3.45	136 593	2.91	281 727	5.37	708 124	3.69
Industrial automation	66 543	3.53	78 431	2.39	78 285	1.92	127 543	2.72	80 282	1.53	431 084	2.25
Transportation	n.a.	n.a.	n.a.		94 870	2.33	239 904	5.11	378 883	7.22	713 657	3.72
Financial services	n.a.	n.a.	n.a.		206 210	5.06	187 921	4.00	266 436	5.08	660 567	3.44
Other services	63 371	3.36	n.a.		282 643	6.94	338 560	7.21	397 145	7.57	1 081 720	5.64
Other manufacturing	n.a.	n.a.	n.a.		757 478	18.60	453 609	9.66	545 985	10.40	1 757 072	9.16
Agriculture	23 315	1.24	35 361	1.08	38 702	0.95	48 026	1.02	27 360	0.52	172 764	0.90
Construction	n.a.	n.a.	n.a.		168 394	4.14	323 363	6.89	240 905	4.59	732 662	3.82
Other	427 803	22.67	1 040 296	31.75	175 523	4.31	107 233	2.128	255 376	4.87	2 006 231	10.46
Total	2 287 378	100	3 276 380	100	4 072 078	100	4 693 831	100	5 249 155	100	19 178 823	100

* Includes: computer systems, computer graphics related, specialised computer turnkey systems, optical scanning and other scanning related, peripherals, computer services, software, voice synthesis/recognition.

Source: Elaboration on EVCA (various years).

number of private venture capital firms entered the market. Thus by the end of 1989 there were 86 venture capital firms in Japan, with a total equity investment of about $2.5 billion.

A significant difference between the US and the Japanese venture capital industry is the low number of start-ups funded by this financial institution in Japan. The substantial lack of a venture capital market and the diffusion of the *keiretsu* form of industrial organisation have in practice hindered the creation of independent start-ups in Japan. A further problem faced by new start-ups is the fact that the Japanese stock market is not designed to handle small start-up companies: therefore venture investment cannot be liquidated, and venture capitalists cannot cash in early-stage financing (see Okimoto, 1985). Moreover the labour market in Japan is such that employees, included R&D personnel, much prefer permanent employment in large established corporations. It is therefore difficult for new, venture-capital-funded companies to attract the most talented people from big corporations.

Nonetheless, although the venture capital industry in the strict sense is still modest in size, and although small industrial companies are not an important vector of innovation, the financial system in Japan shares a number of features in common with the US venture capital system. In both systems there are major company stakeholders with a long-term commitment. These benefit if the companies do well and preserve control rights if the companies do poorly or if the management does not behave in the interest of the stakeholders. Thus a distinction should be drawn between financial activities of the venture capital type, which are very common in Japan, and venture capital in the strict sense.

As far as the financial system is concerned, the most significant difference between the US and Japan is the role played in the latter by the banking system, which offers an illuminating example of the economic impact of the interaction between finance and technology (see Williamson, 1991). The Japanese finance process, therefore, has a set of specific features that render its role and functioning rather different from those in other advanced countries (see Suzuki, 1987). Of outstanding importance among these features is the fact that Japanese banks specialising in medium- and long-term loans to industrial firms act as risk-sharers, since in providing financial backing to industrial investments they in practice become shareholders in the borrowing firm (Mayer, 1988, p. 1181). This means that in the case of Japanese industrial firms the distinction between debt and equity finance is rather difficult to draw, for when industrial firms borrow funds from the banking system they also issue equities that are subscribed to by the same banking system in partial settlement of their loans. The most important component of this financing process is the so-called 'main bank' (see Aoki, 1990).

The Japanese main bank is the largest lender of a particular non-financial company. It owns up to a maximum of 5 per cent of its total stock, and is the manager of the loan consortium that extends long-term credit to the company. Should the company suffer a business crisis, the main bank may reschedule payments and place representatives in top management positions or on the company's board of directors. It may also decide how to reorganise the activity to secure the claims of the loan consortium. In some cases, both the firm and the main bank belong to the same structure, the above-mentioned *keiretsu*, consisting of a group of firms the investment activities of which are financed by a group of affiliated banks and financial institutions (see Hoshi, Kashyap and Scharfstein, 1990). Combinations of banks and financial institutions may thus own more than 5 per cent of stock in non-financial companies: as pointed out by Aoki (1990, p. 14), Japanese financial institutions (including insurance companies) own about 40 per cent of the total stock of listed companies.

In Japan, financial arrangements between industrial firms and main banks undoubtedly resemble those that involve venture capitalists and entrepreneurs in the US. In both cases there is one lead investor that bears most of the risk that the remaining assets will be lost if the firm fails, but receives only a portion of the gains if the firm is successful (see Hansmann and Kraakman, 1992).

The impact of this peculiar role of the banking system on the performance and the financing strategy of the Japanese firm is indirectly confirmed by empirical research carried out by Allen and Mizuno (1989) on the determinants of corporate capital structure among a sample of Japanese firms over the period 1980–3. This study does not provide a clear-cut definition of the Japanese company capital structure, because it reveals both that the most profitable companies fund their investment via internal financing, and that the less profitable companies display instead a lower debt ratio than their more profitable counterparts. According to Allen and Mizuno (1989), these findings reflect the fact that firms' preference for internal finance is affected by the level of taxation, which prevents them from employing retentions in funding their investment outlays.

However an alternative explanation relies on the functioning of the Japanese finance system during the 1980s, when the distinction between ordinary banks and banks providing medium- and long-term loans sharply diminished as a consequence of financial innovation. In this period, as stressed by Suzuki (1987), competition among financial intermediaries brought about an increase in the ordinary banks' activity in the field of medium- and long-term loans and a simultaneous increase in the number of credit banks collecting sight deposits. This apparently irrelevant phenomenon is likely to have significantly affected the corporations' financial structure by encouraging them to resort to external funds and to issue new equities. In fact, by lowering the cost of debt competition in financial markets, an increase in the number of credit banks fostered recourse to external financing and debt, while the ability of Japanese banks to sustain current losses in the expectation of the future compensation represented by equity-type services increased the equities issued by industrial firms in order to settle such debt.

7.3 SUCCESSFUL AND UNSUCCESSFUL VENTURE-CAPITAL-FUNDED START-UPS IN THE UNITED STATES. TWO LITERATURE-BASED CASE STUDIES

As shown in Section 7.2, venture capital has played a much more important role in the USA than in Europe and Japan in the financing of start-ups in the DP industry. Accordingly, the present section describes some significant cases of DP start-ups in that country. These are examples of the relationships between venture capital organisations and DP companies during the mid- to late 1970s and the early 1980s, a period in which venture capital boomed and triggered a major process of new firm formation in regions such as Silicon Valley, the Silicon Prairie and Route 128.

7.3.1 Apple Computer

The introduction of the MITS/Altair computer in January 1985 is conventionally taken as marking the beginning of the personal computer revolution (see Langlois, 1992). However it was probably only in 1977, with the simultaneous introduction of the Apple II, the Commodore PET and the Tandy TRS-80 Model I, that a significant market for the PC industry was created. Thus the name of Apple Computer has been associated with the story of the PC industry since its very beginning, and the company can be taken as a representative case of the way in which venture-capital-funded start-ups can grow rapidly and come to dominate a market. Apple was formed, initially as a simple partnership, on 1 April 1976 by Stephen Wozniak, who had previously worked for Hewlett-Packard, and Steven Jobs, who was on contract for Atari (see Rogers and

Larsen, 1984). It was Wozniack who designed the Apple I, the first product marketed by the newly formed partnership in 1976.

Once the new niche had been identified and partly filled, in the autumn of 1976 the two founders of the company decided it was time to move out of the garage where the first Apple I had been assembled and to create a company with potential for growth. The capital needed to transform a good idea into an industrial activity was raised by Armas C. Markkula – a former marketing manager at Intel – who put $91 000 of his own money into Apple Computer Inc., helped secure a line of credit at the Bank of America, and raised another $600 000 from various venture capitalists.

The Apple II was presented at the First West Coast Computer Fair in spring 1977, and sales immediately began to take off. In the first years, by relying almost exclusively on sales of the Apple II, the company's revenues grew from $750 000 in 1977 to $117 million in 1980. In this initial period, before the introduction of the Macintosh model in January 1984, the success of Apple Computer was due not to technological superiority, but to its provision of real customer support and professional image at a time when this particular market niche was occupied only by Commodore and Tandy.[5] The choice of this successful strategy by Apple Computer was influenced by the same venture capitalists who had helped bankroll the company (see Langlois, 1992).

In this respect it is worth focusing on the various rounds of venture capital financing that led, in December 1980, to the IPO by Apple Computer. This will yield a better understanding of the role played by outside investors in this successful start-up. In effect venture capitalists usually provide at least three critical services in addition to money: (1) building the investor group, (2) helping to formulate business strategies and (3) filling in the management team.

In the case of Apple, this coordination function was performed by Armas Markkula, and the results in terms of the company's total valuation and market performance were excellent. Table 7.7 presents the amount of money raised in the various rounds of financing by both the founders and the venture capitalists.

In the first round of venture capital financing in January 1978 – which coincided with the early success of Apple II – outside investors paid $0.9 per common share, giving the company a fully diluted valuation of $3.063 million. When the initial public offering was made in December 1980 the price per share soared to $22. Since then Apple has been one of the most successful companies in the DP market.

The case of Apple can be taken as representative of the reinterpretation of the Schumpeterian approach presented in Section 2.3 and summarised

Table 7.7 Multiple financing rounds for Apple Computer (1977–80)

	Founders	Founders	Venture 1	Founders	Venture 2	Venture 3	IPO
Date	Mar. 77	Nov.77	Jan.78	Jul.78	Sep.78	Dec.80	Dec.80
Amount raised ($000)	1	115	518	426	704	2 331	101 200
Cumul. funding ($000)	1	116	634	1 060	1 764	4 095	105 295
Stock received	16 640	10 480	5 520	4 736	2 503	2 400	4 600
Total shares outstanding (000)	16 640	27 120	32 640	37 376	39 879	43 306	54 215
% Ownership acquired	100.0	38.6	16.9	12.7	6.3	5.5	8.5
Fully diluted valuation ($000)	1	298	3 063	3 362	11 216	42 061	1 192 730
Price per share ($)	0.00	0.01	0.09	0.09	0.28	0.97	22.00
Estimated ending ownership (%)	30.7	19.3	10.2	8.7	4.6	4.4	8.5

Source: Elaboration on Sahlman (1990).

in Figure 2.1. What we have here, in effect, is an entrepreneurial activity that activated a vector of innovation represented by a 'D'-oriented R&D strategy aimed at developing a product – the microcomputer – the importance of which had been previously misunderstood and underappreciated by established firms. Within this framework, external financing raised through venture capital was the necessary condition for the whole process to unfold. As in the Schumpeterian scheme of Figure 2.1, the diffusion of the new product has engendered a new production pattern characterised by increasing integration between the computer and software industries and a new market structure, with changes in concentration depending on the formation and entry of NTBFs (such as Compaq, Sun, and so on).

7.3.2 Venture Capital in the Development of the Disk Drive Industry

Disk storage devices normally account for about 25 per cent of the cost of a computer system. There are several market segments in the data storage industry, varying from the so-called floppy disk to the high storage capacity hard disk. The first magnetic disks were developed in the second half of the 1950s for mainframe computers and soon began to overtake magnetic tapes in the memory storage market (see IG, 1983). The 14-inch unit became the most popular size, and only after several years was the 8-inch rigid drive introduced, which surpassed the former in unit sales in 1982. In 1980 the first 5.25-inch drives were brought onto the market by Tandon and Seagate, and by 1984 they had outstripped all other drives in terms of total sales.[6] Finally, in 1983 production of the 3.5-inch disk drive was begun even by major suppliers of DP devices (primarily Data Control Co.) and by the end of the decade it had overtaken sales of the 5.25-inch unit.

Thus the explosion in the market for disk drives and related interfaces of the late 1970s and early 1980s reflected not only the growth of demand for PCs, but also more general growth of demand for data storage capabilities in all possible computer applications. However it was the 5.25-inch disk drive that quickly captured a major share of the fast-growing PC market, which came to dominate the overall DP market by the early 1980s. In this period the standard was set by Winchester drives, which rapidly took disk cartridges and disk packs. The disk pack was too expensive and cumbersome compared with the cartridge, which in turn was less dependable and had lower storage density than the Winchester drive.

However the history of the disk drive industry is not entirely a success story. Many of the companies active in the industry around 1980 failed as a consequence of the intense rivalry that followed their rapid entry into the market (see Sahlman, 1990). In the years between 1977 and 1984, venture capital firms invested approximately $588 million in 73 manufacturers of disk drives (see Table 7.8). Of these, 17 were manufacturers of floppy disks, 43 of Winchester disks, seven of optical disks and six of other disk-related products. In general, the expansion of the disk drives market was characterised by a very strong process of new firm formation fostered by the availability of venture capital financing. However, as the Table 7.8 clearly shows, the 'Golden Age' for Winchester disks were the years between 1980 and 1983, whereas by 1984 this market had begun to shrink.

Priam Corporation can be taken as an enlightening example of both the function of venture capital and outside financing in the development of this crucial branch of the DP industry when the PC revolution took place, and the problems faced by disk drive manufacturers unable to cope with technological change.[7] Priam was founded in 1978 by Bill Schroeder and Alonzo Wilson, two former employees of Memorex Corporation, a large manufacturer of computer storage devices. In its early years, Priam served the market of mid-range computers, which was characterised by demand

Table 7.8   Investments of VC firms in disk drive companies by disk drive technology (1977–84)

	Floppy disks	*Winchester disks*	*Optical disks*	*Other disks*	*Total*
1977 (no.)	1	2	0	0	3
($)	200	780	0	0	980
1978 (no.)	6	3	0	0	9
($)	5 955	6 383	0	0	12 228
1979 (no.)	3	6	0	0	9
($)	8 250	5 220	0	0	13 470
1980 (no.)	7	5	0	0	12
($)	19 766	11 100	0	0	30 866
1981 (no.)	9	12	0	0	21
($)	23 414	33 254	0	0	56 668
1982 (no.)	8	19	1	1	4 000
($)	30 862	67 540	150	4 000	102 552
1983 (no.)	7	22	1	2	32
($)	29 125	184 356	9 326	11 000	233 807
1984 (no.)	5	20	6	5	36
($)	16 915	83 480	15 162	21 938	137 495
Cumulative (no.)					
rounds ($)	46	89	8	8	151
	134 487	392 113	24 638	36 938	588 176

Source: Elaboration on Sahlman and Stevenson (1985).

for devices larger and more reliable than those required in the personal computer market. After leaving Memorex, Schroeder and Wilson contacted a group of venture capitalists, from whom they raised more than $14 million in a series of sales of convertible preferred stock. These funds represented 60.5 per cent of the total funds raised by Priam from its inception to June 1983, when the IPO of its shares was made.

During the first four years of activity, total revenues at Priam were approximately $36 million and the company spent $7 million on R&D and product development. Soon after its entry in the market, Priam negotiated the first of a series of several 'leaseline' agreements with the Bank of America, which enabled it to obtain the equipment and buildings that it needed to begin and sustain business. Thus, like the Japanese industrial firms outlined in Section 7.2.3, Priam relied upon long-term credit provided by a sole lender to strengthen its production capacity. However, high interest rates on short-term loans and a lack of confidence by venture capital companies in Priam's expected returns in 1982 forced Schroeder to resort to a new source of financing, represented by a number of major insurance companies. These companies invested $7 million in Priam by acquiring a package of securities consisting of approximately equal dollar amounts of convertible subordinated debentures and a new series of convertible preferred stock.

In 1983 the financial crisis compelled Priam to cut back on its long-term R&D, in particular a project to develop the company's capacity in 5.25-inch disk drives. Since long-term damage had been done to the product development effort, Schroeder and his board considered purchasing another company, Vertex Peripherals, which at the end of 1982, together with Micropolis Co., had been the first company to introduce a high capacity, high performance 5.25-inch disk drive. Vertex also faced financial problems and was seeking the funds needed to build a new factory. Thus, by merging with Vertex, Priam would have been able to overcome its R&D problems and gain a competitive edge over other manufacturers of disk drives. The alliance would have also benefited Vertex in its effort to enhance its production levels. Unfortunately Priam did not formally present its offer to the Vertex board of directors, and Vertex issued new convertible preferred stock in order to raise the funds it was seeking. In spite of a drop in the share prices of disk drive companies, private investors took enthusiastically to the Vertex offer. Thus the possibility of a merger with Priam was no longer considered.

This story demonstrates the closeness of the relationship between financial and technological strategies for companies competing in the dynamic peripherals market during the early 1980s. Having lost a good chance to enhance its technological capability through a merger with Vertex, Priam was caught in a downward spiral from which it never escaped. The company filed for bankruptcy in 1989.

7.4 CHANGING STRATEGIES: FROM THE NEW TECHNOLOGY-BASED FIRM TO THE ESTABLISHED HIGH-TECHNOLOGY FIRM

Since venture capital is the most appropriate source of early-stage financing for high-technology start-ups, it was argued in Chapter 4 that 'loyal' top executives in high-technology PEPFs issue new equity whenever they decide to undertake *aggressive* R&D projects that are likely to give the company a competitive edge and improve its market share and economic performance. On the basis of official data, it is difficult to identify for each company the technological features and complexity of different R&D projects: accordingly the R&D projects undertaken by one certain company may be more 'D'-oriented than those of another. Thus the theoretical assumption of Chapter 4 that equity financing is preferred by companies pursuing 'aggressive' R&D strategies cannot be reliably tested by using information derived from the companies' annual reports. Empirical investigation must consequently take a much broader perspective and assume that an increase in shareholders' equity and total debt in period $t+n$ is accompanied by an increase in R&D expenditures during the same period.

Under this assumption, a simple equation can be estimated in order to capture the relationship between the rates of growth of shareholders' equity and long-term debt, and the company's R&D activity in general. The model to be estimated is

$$\Delta R\&D_t = f(\Delta SE_t, \Delta LTD_t) \quad \text{with } f_1, f_2 > 0 \tag{7.1}$$

where $\Delta R\&D_{t+n}$ is the actual growth rate of the company's R&D expenditures, ΔSE_t is the growth rate of the company's shareholders' equity, and ΔLTD_t represents the growth rate of long-term debt.

This simple model has been tested against a sample composed by the large companies listed in the *Macmillan Directory of Multinationals* (Stafford and Purkis, 1989) which, according to the *Datamation 100* rankings, earn more than 40 per cent of their total revenues from sales of DP and related products. Using this procedure it was possible to single out eighteen multinational companies[8] displaying the requisite features, and to consider the relevant data for the period 1983–7. The choice of the period is consistent with the fact (stressed in Chapter 6) that around the mid-1980s the elasticity of demand for DP equipment to improvements in technology was still high, whereas it started to decline in subsequent years. The estimated equation is thus the following:

$$R\dot{\&}D_t = \beta_0 + \beta_1 \dot{S}E_t + \beta_2 L\dot{T}D_t + u_{it} \tag{7.2}$$

where u_{it} is the error term, $i = 1, 2, \ldots, 18$ stands for the company and t is the relevant year.

Using Equation 7.2 it was then possible to estimate four annual OLS cross-company regressions, from 1984 to 1987. However, to obtain more efficient estimates I tested the possibility of working with 72 observations by pooling cross-section and time series data. The hypothesis that both slope coefficients and the intercept of Equation 7.2 are constant over time has been accepted by the F-test (see Judge *et al.*, 1980). Accordingly, Table 7.9 presents the results arising from the set of pooled regressions carried out, with $R\&D$ as the dependent variable, using the pooled ordinary least squares estimator (POLS). The specification turned out to be exempt from heteroscedasticity. $\dot{S}E$ and $L\dot{T}D$ are alternatively excluded from regressions 7.2.2 and 7.3.3.

In the three regressions, the coefficients of the independent variables are significant and have the expected sign. The low value of $\bar{R}^2$ is presumably due to a high level of company heterogeneity and to the aggregated nature of the available R&D data which, as stressed above, do not allow for any distinction between different R&D strategies. However these results are

Table 7.9 Pooled cross-section, time series regressions[1]

	7.2.1	7.2.2	7.2.3
Intercept	0.128	0.136	0.145
	(9.249)	(9.613)	(11.150)
$\dot{S}E$	0.181	0.190	
	(2.782)	(2.788)	
$L\dot{T}D$	0.032		0.033
	(2.963)		(2.970)
$\bar{R}^2$	0.178	0.087	0.099
F	8.706[2]	7.771[2]	8.821[2]

Notes:
1. *t*-statistics in brackets (all significant at 0.005 level).
2. Significant at 0.01 level.

satisfactory for the purposes of the analysis conducted in this book, because they provide statistical support for the assertion that the choice of the source of financing significantly affects the patterns of innovative activities. They show in particular that in large managerial companies belonging to a high-tech industry the issue of new equity is an important source of the funds spent on R&D. This finding is not only consistent with the theoretical framework developed in Section 4.4. It also confirms Schumpeter's (1928) suggestion[9] that, with 'trustified' capitalism, the working of the financial system enables economic agents to evade budget constraints.

7.5 TAKEOVERS, COOPERATION AGREEMENTS AND THE PATTERNS OF INNOVATION IN THE DP INDUSTRY

In-house formal and informal R&D is not the sole source of innovation for established companies belonging to advanced industries. In effect, R&D partnerships, agreements and strategic alliances are valuable alternative ways to gain a competitive edge in technological capability. They take the form of vertical and horizontal bilateral contracts, the aim of which is to handle the problems arising from the uncertainty of innovative processes and from the indivisibilities, inappropriability, asset specificity and tacitness that can provoke organisational failures (see Teece, 1989). In the case of DP and other high technology industries reliant upon internal R&D capability, acquisitions have been a rapid means to introduce further innovations (Link, 1988), and bilateral contracts of this kind have assumed increasing importance. The same is the case among informal R&D partnerships, even among rival companies (see von Hippel, 1988, 1989).

However, a view widely advanced in the economics literature is that a wave of mergers, acquisitions and leveraged buyouts distracts managers from long-term decisions, including R&D investment. It has thus become commonplace to regard takeovers and R&D as substitutes, in the sense that they represent two alternative ways of acquiring knowledge capital. In effect, a company can acquire knowledge capital either by internally investing in R&D programmes, or by purchasing another firm after its R&D programme has yielded successful results. However the empirical evidence is controversial and does not indisputably support this interpretation.

In a study of the R&D performance of US manufacturing firms, Hall (1988) demonstrated that there seems to be some significant increase in R&D intensity in computers and electronics by the acquiring company side around the time of acquisition. This finding perhaps indicates that DP companies engaged in acquisition activity should invest more rather than less in R&D if they are to exploit the value of their acquisitions in full. Moreover the same study showed that in some cases the R&D intensity of acquired firms was higher than that of the firms that took them over; a result that suggests that under certain circumstances firms do indeed acquire knowledge capital by purchasing other firms.

Within the DP industry as a whole in the 1980s and early 1990s, mergers, acquisitions, management buyouts (henceforth MBOs)[10] and cooperation agreements represented a major factor in firms' external growth; and in some cases they led to the undertaking of innovative projects. In particular, R&D joint ventures and consortia (see Evan and Oik, 1990) focusing on the 'D' rather than the 'R' represent a cost-sharing strategy that can be usefully pursued by companies facing financial constraints. Therefore most strategic alliances depend closely on the functioning of the capital markets, and in many cases they provide industrial companies with the means to raise the knowledge capital they require to implement their projects.

According to data provided by Grasso (1989), the number of cooperation agreements and takeovers grew significantly between 1980 and 1988. This process of integration was driven by non-equity agreements, whereas a secondary role was played by MBOs. The figure apparently changed between 1988 and 1993, when acquisitions and mergers overtook cooperation and other non-equity agreements (Table 7.10). Of great interest is the figure on cooperation agreements, since these gave rise in several cases to partnerships between hardware manufacturers

Table 7.10   Main mergers, acquisitions, cooperation agreements and management buyouts in the DP industry (1988–92)

	Type					
	M	*A*	*JV*	*CA**	*MBOs*	*Total*
1988	2	7	3	7	1	20
1989	6	8	5	10	–	29
1990	5	12	–	9	1	27
1991	4	14	–	6	–	24
1992	6	8	4	7	–	25
1993	2	8	7	7	–	24

* Includes: consortia, exchange of patents, mixed shareholdings, technical and trade agreements.
Source: Elaboration on Assinform (various years).

and software or service suppliers. Among such agreements, especial mention should be made of those resulting in 1988 from the joint effort by AT&T and 30 allies in the creation of the UNIX International Inc., by IBM and others in the creation of the OSF (Open Systems Foundation) consortium,[11] and of Compaq and 70 other PC vendors in the creation of the EISA (Extended Industry Standard Architecture) consortium.

Inspection of takeover and cooperation agreements by technological area within the DP industry (Table 7.11) shows that only between 1988 and 1990 did EDP equipment represent the main field of takeovers and cooperation agreements, whereas after 1991 the area of software, services and image processing began to comprise the highest number of such operations (50 per cent of total operations in 1993).

Thus – reinforcing a tendency that has emerged since the mid-1980s (see Grasso, 1989) – the process of integration among DP companies has increasingly focused on the intangible resources of partners, relegating the core technology to a secondary role.

One possible explanation for the decreasing importance of takeovers and cooperation agreements in the EDP field is the role of platforms such as OSF, UNIX, and EISA in determining intrasectoral equilibria and

Table 7.11   Main mergers, acquisitions, joint ventures, cooperation agreements and MBOs in the DP industry (1988–92)

	Technological area				
	EDP equipment[1]	*Office equipment*[2]	*Telecom equipment*[3]	*Software equipment*[4]	*Total*
1988	12	1	1	6	20
1989	15	2	4	8	29
1990	14	–	4	7	25
1991	7	2	6	9	24
1992	8	–	5	12	25
1993	4	2	6	12	24

Notes:
1. Personal computers, workstations, palmtops, minicomputers, mainframes, supercomputers, superminis, peripherals, terminals, printers, floppy disks, smart cards, CD WORMs.
2. Office automation in general (banking automation, large systems, system integration).
3. TLC, cellular phones.
4. Software, services, image processing.
Source: Elaboration on Assinform (various years).

directing competition among the leading companies in this market (see Francis, 1989).

Conversely, the area of software, services and maintenance is still a dynamic market in which companies battle for leadership. Evidently in this case, takeovers and cooperation agreements are perceived as a way of gaining a competitive edge, and they are also a strategy pursued by those companies (such as IBM, Digital Equipment and Fujitsu) that have their core business in hardware rather than in software.

7.6 SUMMARY

This chapter has presented empirical evidence on the financial structure and the innovative activity of companies belonging to the DP industry during periods characterised by a significant process of technological change.

We have seen that venture capital significantly boosted the process of technological change in this industry between 1975 and the early 1980s by allowing new, fast-growing microcomputer and peripherals manufacturers to enter the market. However this important source of funds appears to be sufficiently well developed only in the UK and the USA, whereas in continental Europe it is still in its infancy. Different again is the case of Japan, where the entire financial system is able to channel risk capital towards the exploration of new technological possibilities through the 'main bank' system.

Moreover, around the mid-1980s large, well-established DP companies appeared to rely on equity financing as an important source of funds for their R&D investment. This and the above findings are consistent with the theoretical assumptions presented in Chapter 4, and confirm the main argument of the present work, which is, that the firm's financial structure *does* matter as a determinant of innovative investment decisions.

Part III

The Role of Government

8 Public Policies towards the Financing of Innovative Activities

8.1 PUBLIC POLICY AND THE DEVELOPMENT OF NEW TECHNOLOGIES

The historical evidence presented in Chapter 5 and the empirical findings of Chapters 6 and 7 demonstrate that the features of financial institutions and the firm's ability to overcome budget constraints are of great importance in fostering the formation of the diffusion cluster, which permits, through the assimilation of a new technological style, radical reform of the techno-economic subsystem. In this connection it is therefore possible to make a number of suggestions regarding the policy interventions most likely to favour investments in 'D'-oriented R&D activities and which may directly and indirectly encourage the wide diffusion of new technological styles. These suggestions are intended to apply to those countries that take science and technology as a primary goal of public policy. If it is true that (as suggested in Section 5.3.5) the current period is one of transition from a downswing to an upswing in the periodisation of Kondratieff long waves, science and technology policy should today more than ever serve the purpose of promoting and facilitating those investments in high-tech activities that might shorten the transition from an exhausted to a new technological style.

Accordingly, in the first part of this chapter (Section 8.2) a comparison is made among public policies intended to promote various sources of innovation financing in the USA, Europe and Japan. In the light of such experiences and of the results of the previous chapters, some suggestions are then made regarding (mostly microeconomic) innovation policies specifically designed to channel financial resources towards innovative investments in an efficient manner (Section 8.3).

8.2 INTERNATIONAL COMPARISONS: USA, EUROPE AND JAPAN

From the evidence of previous chapters, one may suggest two general strategies in innovation policy: (1) in the case of small high-tech start-ups

(NTBFs), public policy should pursue a series of different goals, ranging from the creation of technology-sharing cooperative agreements to the provision of incentive schemes for venture capital firms; and (2) in the case of established publicly traded companies (PEPFs), innovation policy should instead focus on two main issues: (a) the introduction of laws regulating corporate dividend decisions (fiscal policy) and (b) the provision of incentive schemes aimed at promoting technological competition among oligopolists.

From a broader perspective, it of course remains true that new technological styles can be typified by public funding to universities and public research centres. In effect these institutions perform a crucial function in the development of the general and basic knowledge, exploitable at the industry level, that is necessary for the crystallisation and subsequent diffusion of new technological styles in the techno-economic subsystem.

As regards recent experience, in advanced countries innovation policy is usually perceived as a branch of competition policy. On the assumption that the optimal innovation policy is one that increases civilian investments in R&D and encourages the adoption of innovative production technologies, from a financial viewpoint current innovation policy in most countries is a mix of direct subsidies for small firms and new entrants and tax-based incentive schemes for large established companies. Nevertheless the problem of diverting financial resources to R&D has been addressed differently, at least in terms of the resources involved, by each of the most advanced countries. The share of national R&D that is privately financed differs significantly from one country to another: among the G7,[1] for example, the share of R&D financed by the business enterprise sector between 1975 and 1991 was extremely high in Japan (more than 70 per cent on average), whereas it was markedly lower in the other six countries (indeed, less than 40 per cent on average in Canada). This evidence of course has implications for policies related to the financing of innovation: in designing their public policies, countries in which company-financed R&D intensity has been lower than in others should introduce (and in some cases have in fact introduced) extra stimuli to generate autonomous innovative investments.

In consideration of intercountry differences, before providing some concluding policy suggestions it is therefore worth summarising the aims and features of R&D financing policies in the USA, Europe and Japan.

8.2.1   USA

In the USA – leaving aside the government function in the financing of basic research carried out either by universities or by industry – the two policy issues corresponding most closely to the arguments set out in the present book are antitrust policies towards cooperation agreements in R&D (see Link and Tassey, 1989; Jorde and Teece, 1992; Geroski, 1993) and R&D tax policies (see Mansfield, 1986; Hall, 1993).

American policy makers have usually taken rivalry and the absence of artificial barriers to entry as the most efficient means of obtaining optimal economic performance. Accordingly, US antitrust laws do not apply to agreements regarding basic research, and they permit with some restrictions those that relate to applied research provided they promote innovative processes that would not otherwise take place (see Martin, 1993). The arguments in support of this permissive attitude turn on the existence of externalities of various types: technological spillovers, pecuniary externalities, environmental externalities (see Geroski, 1993).

Technological spillovers are likely to arise because firms involved in cooperative R&D tend to supply each other with information produced in the undertaking of their respective activities. In this way, each firm involved in a given R&D agreement is likely to benefit not only from the results of the cooperative project itself but also from the further information that becomes available, without having to pay for it. Pecuniary, positive externalities occur as a result of the risk-sharing purpose of most cooperative R&D: the action of firm i involved in the R&D project affects the competitive position of the other firm j. Finally, environmental externalities arise when cooperation in R&D influences the partners' ability to cooperate on other lines of activity and gives rise to broader non-R&D-related commitments among them.

However, in spite of its permissive attitude, the National Cooperative Research Act (henceforth NCRA) of 1984 precludes those 'D'-oriented R&D cooperative agreements that also entail post-innovation cooperation. Irrespective of their share of the total market, firms are therefore prohibited from signing agreements that include joint commercialisation following innovation. The NCRA offers a shelter from antitrust only for R&D activities, provided that these have already been registered. Its importance relies on the fact that it is the first piece of US legislation to mitigate the traditional antitrust treatment of innovation.

Nonetheless the restrictive rules of the NCRA have been increasingly subject to criticism by American students of technology (see, for example, Jorde and Teece, 1992), who contend that the joint manufacturing and production of innovative products is in some cases necessary to stimulate further innovations and make joint R&D profitable. They suggest that interfirm agreements of this kind should be permitted between firms with less than 20–25 per cent of the relevant market; firms that otherwise would be unable to commercialise the results of their R&D activities successfully.

R&D tax policy in the USA has three main components (Hall, 1993): (1) the writing off of most R&D expenditures against corporate income for tax purposes; (2) an R&E (research and experimentation) tax credit computed by taking qualified R&D expenditures in excess of a given level, multiplying this amount by the statutory credit rate, and then deducting it from corporate income taxes and (3) a foreign source income allocation rule that grants US multinationals credits on taxes due on the foreign source income when such income is allocated to R&D.[2]

Broadly speaking, in the USA both antitrust policy on cooperation agreements in R&D and R&D tax policy provide strong incentives for company-financed R&D. These policies are therefore effective only on the assumption that US companies fund their R&D investments out of retentions and that banks and other financial institutions are in any case able to provide the needed capital. In this connection, the size of financial markets and institutional investors and their attitude towards risk are crucial variables.

8.2.2   Europe

The main aim of innovation policies in Europe has traditionally been to strengthen the industrial potential for R&D in order to meet the 'American Challenge' (as defined by Servan-Schreiber, 1965) during the 1960s and the 1970s, and the 'Japanese Challenge' in the 1980s. This target was pursued by various national governments through the promotion, mainly via mergers and subsidies, of large firms in technologically strategic industries. A shift in the focus of technology policies from national governments to European Union (EU) institutions occurred in the 1980s, with the joint development of interventionist programmes intended to nurture new high-tech industries and diffusion-oriented schemes. In particular, the EU has established programmes designed to promote general research, even at the precompetitive stage, and cooperative research. From this perspective, the field of data processing has benefited from the ESPRIT and RACE programmes, which focus on the adoption of certain key innovations, and therefore encourage the formation of a modern industrial structure exploiting the most advanced production technologies.

As regards cooperation agreements in R&D, the European Commission permits this kind of strategic alliance only 'where the know-how resulting from the common R&D contributes substantially to technical or economic progress and constitutes a decisive element for the manufacturing of the new or improved products' (EC Commission, 1985, p. 38). Thus cooperative R&D – unless it is included in special EU programmes – is permitted on a limited scale and only if there is evidence that it will ultimately benefit consumers through product innovation. Smaller firms, however, fall outside these restrictions, and receive a wide range of incentives on a national or even regional scale. In Italy for example, Law 317 of 1991 grants financial assistance to cooperative R&D projects involving industrial firms with fewer than 200 employees and private or public research centres.

Tax incentives and deductions for R&D investments are regulated mainly on a national basis, although of course they cannot be in breach of the EU rules on competition (see EC Commission, 1986). Thus, in most EU countries, enterprises can reduce their taxable profits by a deduction (in general ranging between 20 per cent and 50 per cent) for R&D investments that can be either immediate or deferred. Tax incentives are also granted in order to stimulate venture capital activities with a view to the creation of new high-technology enterprises. In such cases (for example in Belgium, France and The Netherlands) the fiscal incentives benefit all parties involved in the research activity, from the high-tech firm down to the venturing firm and even private investors.

There seems to be a favourable attitude in Europe towards the provision of incentives for the external and internal financing of innovative activities. However a major problem affecting all EU countries with the exception of the UK is the inadequacy of the stock market, which is not designed to handle small start-up companies and is thus in itself an obstacle to the widespread diffusion of venture capital. As shown in Chapter 7, standard R&D partnerships provide outside investors of the venture capital type with the option of receiving stock in the company as soon as sales of the new product commence. Therefore, if the stock market prevents venture capital professionals from cashing in their investments, financial institutions of the venture capital type cannot play any major function in channelling resources towards innovative start-ups.

8.2.3 Japan

The case of Japan is somewhat peculiar in comparison with those of the USA and the EU, for at least two reasons: (1) bank finance is very pronounced in that country – Japanese firms therefore encounter no difficulties in using bank credit to fund their R&D activities (see Section 7.2.3 above),[3] and (2) Japanese investors are on average highly risk-averse and thus have strong preferences for investment in safe assets (bank deposits and postal savings) rather than equity investment.

The wide diffusion of bank financing in Japan notwithstanding, joint R&D effort has been encouraged by the Japanese government, which has introduced tax incentives of various types in order to induce firms to invest greater sums in R&D. Nonetheless the level of government direct R&D subsidies is traditionally low in this country (see Freeman, 1987). The emphasis has instead been on other positive incentives such as pragmatic antitrust enforcement, creation of a favourable macroeconomic environment, and consensus building across public and private sectors (see Okimoto, 1985).

With respect to joint R&D, the Japanese government has promoted cooperative national research projects in the form of joint research associations operating in areas of high risk and heavy cost, but of potentially great commercial benefit, in order to enhance the level of basic knowledge. In the DP field, these include the computer project (1972–6), the software development project (1976–81), and the very large-scale integration (VLSI) project (1976–81). The latter, however – like the more recent fifth generation computer project – failed to push Japanese DP and semiconductor technologies beyond the frontiers of knowledge.

Continuous and cooperative relationships in design and R&D have simultaneously taken place between manufacturers of parts and equipment, without this entailing any kind of vertical integration or post-innovation collusion. In the case of the DP industry, market-oriented cooperative research carried out by private companies has greatly benefited from more basic research undertaken in the laboratories of the Nippon Telephone and Telegraph Company (NTT). At least until the mid-1980s, this semi-public organisation (now partly privatised) directly carried out or funded around 20 per cent of total Japanese computer R&D and transferred its results to private companies (see Flamm, 1987). Some of this R&D activity has been undertaken by NTT through research partnerships with the participation of other companies.

All these institutional frameworks pursue the goal of disseminating R&D output while ensuring rivalry in the exploitation of the commercial results of R&D: in contrast with the US, R&D cooperation in Japan *does* entail post-innovation competition among participating firms and is designed to improve market performance and benefit consumers, rather than increase profits. In practice, the undertaking of cooperative R&D in Japan serves the purpose of developing a broad base of fundamental technologies for autonomous use by the member companies.

Tax incentives are also used in Japan to enhance the level of civilian R&D investments. For example, R&D spending above the highest previous level is eligible for a 20 per cent tax credit, the purpose being to persuade firms to invest a larger amount of money every year in research activities.

The Japanese government's attitude to cooperative R&D and R&D tax policy reflects the position of the Ministry of International Trade and Industry (MITI) as regards programmes aimed at creating a commercially competitive industry. MITI, in fact, has traditionally focused on applied rather than basic research in order to strengthen Japanese industrial performance, then using industrial policy to foster commercial applications of R&D activities carried out by home industries.[4] As a consequence the provision of seed or start-up capital to high-tech, newly established companies has never been an important component of MITI's programmes, because of its strong preference for R&D projects with a quick technological payoff.

8.3 SOME POLICY SUGGESTIONS

A significant process of new firm formation in high-tech industries is a guarantee against the creation of oligopolistic market structures, which, in some cases, may increase private profits without benefiting consumers. As shown by the US experience, this process relies on the availability of risk capital and consequently requires a developed venture capital market. Public policy in countries without a developed high-tech industrial sector should therefore provide incentives for the creation of a venture capital market, and they should organize the stock market so that it can handle small start-up companies that wish to go public and allow venture capitalists to cash in their early-rounds financing.

With regard to the financing of innovative start-ups, venture capitalists may find the opportunity to participate in the development of new technologies attractive for several reasons: tax shelters, for example, or such benefits as potential equity position, earlier payout, and no dependence on profits (see Spohr and Wat, 1985). All these aspects are closely related to public policy orientation towards venture capital. In effect, tax shelters enable outside investors to deduct a significant portion of their investment against their ordinary income, and may even account for 80 per cent of the investment. Potential equity position requires ease in going public for the new company, so that investors can be attracted by the opportunity to receive stock in the company once the initial public offering has been made. Earlier payout can be obtained through royalty R&D partnerships, which represent an attractive form of investment for venture capitalists. In this case investors begin to receive cash returns as soon as sales of the new product commence, irrespective of the company's profitability. Since independence of profits insulates outside investors against the uncertainty of the long-term profitability of the new company, incentives to royalty arrangements are an important policy instrument in all those economic systems (such as, for example, the Italian and the Japanese ones) that are characterised by highly risk-averse investors.

Because small high-tech firms are a deterrent against the formation of oligopolistic market structures, cooperative R&D enables such firms to exploit scale economies in research activities and thus enhance their innovative capability. Promotion of cooperative R&D among small firms requires the introduction of incentive schemes that give financial backing for alliances created to undertake research activities whose results will be shared among the participants. Pursuit of this goal through policy intervention, enables small firms to reach the minimum efficient scale to undertake uncertain R&D projects and to compete technologically against larger firms. If allowed to exploit scale economies in R&D, and if they are able to obtain the necessary financial support, small high-tech firms will probably continue to occupy an important role in the overall process of technological innovation. In particular, they will be able to explore those natural trajectories of technology that large established companies find less attractive, thereby contributing to the fuller exploitation of each technological style.

A risk-shifting support to small firms should not be restricted to R&D activities alone: financial constraints often prevent the adoption of improved machinery with embodied technological change even in sectors that are users rather than producers of R&D. Thus, financial support for the adoption of new production technologies is important in the case of small firms belonging to 'traditional' sectors of manufacturing: firms that tend to resort mainly to technological change embodied in new machinery and capital equipment as a source of innovation (see Santarelli and Sterlacchini, 1994b). Through the adoption of innovative machinery and capital equipment, such firms may develop further innovations, even of the incremental type, that are likely to fulfil an important function in the creation of a national system of innovation.

However, also important for the development of innovations are large established companies. In this connection, the financing of aggressive R&D strategies should be promoted through corporate laws that affect corporate financing decisions, and through tax incentives for those oligopolistic firms more reluctant to undertake innovative activities.

The problem to be overcome with respect to large established companies (PEPFs) is management preference for the internal financing of investment projects. This enables them to avoid payment of larger dividends to the new shareholders, and to enjoy, together with incumbent shareholders, the tax benefits deriving from the reinvestment of the company's earnings (see Fox, 1987). However, aggressive R&D strategies are usually very expensive and they cannot be undertaken without resorting to the issue of new securities for cash or to debt financing. The choice between equity and debt financing is in this case rather controversial: the large company prefers debt financing because payment to debt holders reduces its taxable income, whereas outside investors prefer to hold equity rather than debt since the former usually has tax advantages over the latter. In this situation, a simple readdressing of fiscal policy is advisable: corporate tax should be reduced in order to reduce the preference for external over internal financing and the use of debt instead of equity when external financing is sought. In this way, each shareholder could be taxed on the basis of the earnings accruing from her/his shares. Tax benefits moreover do not affect management investment decisions. In a system of this kind, the PEPF is more likely to use equity financing when undertaking uncertain R&D investments, for at least two reasons: on the one hand equity financing is a way of sharing risk with external investors; on the other, it does not reduce the level of post-investment earnings per share. In effect, an innovative investment of the kind considered here is intended to enhance the company's total earnings. Consequently, should it prove successful, all the stockholders will benefit from the increase in the company's value and return stream. Conversely the company is likely to use internal or even debt financing when undertaking routine investments that do not affect the level of its total earnings.

In order to convince reluctant oligopolists to invest more heavily in innovative activities, tax-based incentive schemes are preferable to government direct subsidies. In Japan and France, for example, large companies may be granted a research allowance that permits them to deduct from taxable income an amount equal to a given percentage of the increase in their total R&D expenditures with respect to the previous fiscal year. However such tax-based incentives should preferably be provided to oligopolistic companies only, since in different conditions they may create a tendency to over-invest in R&D by artificially fostering the process of technological competition (see Dasgupta and Stiglitz, 1980).[5]

If the aim of policy intervention is to channel the largest possible amount of financial resources towards innovative activities, carried out either by start-up or established companies, information services should be introduced to give advice to potential investors. Of course this also entails companies being willing to disclose information on the goals of their R&D projects, and on their organisational and financial structure. A government-sponsored information service should thus collect and disseminate information about both research projects and the supply of financial resources, thereby helping investors and companies who are starting R&D activities to make their decisions and choices. In effect, the working of the finance process and the results of inventive activity also depend on the information available to both potential investors and companies seeking the financial resources necessary to implement their research efforts.

The above policy recommendations refer to an international economic community in which the various regions converge in their R&D-related financial structures. Yet the historical evidence presented in Chapter 5, and the comparisons between the US, European and Japanese financial systems conducted in Chapter 7, reveal that even the most developed countries apparently have a limited ability to converge. However the recommendations by leading economists in the USA for greater downstream cooperation in R&D and product development, as well as the attempt to establish a developed venture capital market in continental Europe, confirm that there is as much convergence as the prevalence of the different financial styles permits. This convergence will probably be even more marked if the ongoing processes of integration among European (EU) and North-Atlantic (NAFTA) countries leads to the greater internationalisation of the various national economies and to a consequent reduction in regional differences.

9 Conclusions and Synthesis of Results

In this book I have used different methodological perspectives to analyse the interaction between innovations and significant events in the financial system on the one hand, and the development and diffusion of technological innovations on the other. In following this line of inquiry I have developed two main arguments.

First, from a macroeconomic perspective, the assimilation of new technological styles into the techno-economic subsystem takes place in concomitance with a change in the institutional arrangements characterising the financial system. Although financial and technological innovations occur in the first instance as autonomous events, periods of economic growth and structural change are punctuated by financial innovations adopted to finance the diffusion of a new technological style, because previous methods of finance have proved inadequate.

Second, from the innovating firm's perspective, the choice of finance structure *does* matter in determining the choice among alternative, 'D'-oriented R&D projects. In effect, the innovating firm is better informed than the market about the possible return stream generated by its specific investment in innovative activities and it can therefore choose the finance structure best suited to attract investors.

Analysis has been conducted in theoretical, historical and empirical terms. Theoretically, Chapter 2 reconsidered the original contribution of Schumpeter in order to include *invention* among the most important economic variables. Schumpeter's insights can be employed more appropriately to identify, on the one hand, the long-run relationships between financial and technological variables and their impact upon cyclical phases of economic development; and on the other, the way in which firms fund their innovative investments. These two lines of analysis were examined closely and expanded in Chapters 3 and 4 respectively.

In Chapter 3 some original insights of evolutionary theory were used to introduce the concept of financial style. Just as technological styles are used to denote new matrices of knowledge exploitable in production activities, so financial styles denote the characteristic features of financial instruments and institutions in the presence of well-defined interests, 'moral' attitudes and overall economic pressures and tendencies. Therefore the emergence of a new financial style typifies the transition from a socio-institutional framework to another, which, in turn, is of particular importance in permitting the emergence of a new techno-economic subsystem, that is, a quantum leap in potential productivity. Since they denote a reform of the socio-institutional framework, financial innovations set out the institutional arrangements most akin to the configuration that the techno-economic subsystem is expected to assume after the emergence of a new technological style.

In Chapter 4 modern finance theory was used to describe the financial strategies of both innovating small firms (NTBFs) and large corporations (PEPFs). It was shown that (a) the finance structure of the firm *does* matter in determining the choice among alternative investment (R&D) projects, and (b) that small firms and large firms resort to different sources of financing when undertaking aggressive R&D strategies. As regards point (a), the analysis considered how NTBFs may enter into new long-term financing contracts (such as venture capital) only in order to fund R&D projects aimed at obtaining specific innovations. These firms are assumed to be unable to employ their retentions for the funding of innovative activity because the 'D'-oriented R&D project they wish to undertake requires an investment of funds in excess of their resources. As regards point (b), the analysis concerned the strategy adopted by PEPFs to attract outside investors in order to raise the funds needed to implement their aggressive R&D activities. It was shown that PEPFs usually allocate a limited amount of money every year to different projects, and that their demand for funding is thus likely to exceed the amount set aside for R&D purposes in any particular year whenever they decide to undertake a 'D'-oriented R&D project. By issuing new equity they are therefore able to gather fresh funds, which can be used to implement the innovative investment decision.

From the historical, quantitative analysis conducted in Chapter 5, it then emerged that since the implications of Schumpeter were empirically sound, financial innovation has played a crucial role in successive phases of economic development. Comparison of mean values in the *distribution* of financial and technological innovations during the sequence of upswing and downswing periods identified by various students of Kondratieff long waves confirmed that:

1. The financial component of the overall socio-institutional framework is more likely to undergo change and reform during upswings than downswings.
2. The number of basic technological innovations tends to be higher during downswing periods, when the search for technological

innovations is more likely to move in directions that are different from those entailed by the obsolescent technological style.

The qualitative analysis conducted in the same chapter shed light on the relationship among the five groups of technological and financial styles that have characterised the five development cycles of the Kondratieff type since the 1790s. The analysis focused in particular on the following: the interaction between the exploitation of water power and stationary steam power on the technological side and the use of merchant capital in industrial activities on the financial one; that between the railways technological style and the stock exchange financial style; that of the chemicals, electricity and steel technological styles with the adoption of the international gold standard; that between the 'Fordist' technological style and the financial style based on the crucial function exerted by central banks in guaranteeing monetary and financial stability; and, finally, the role of non-banking financial institutions such as venture capital firms and pension funds in the early diffusion of information technology and the biotechnology industries.

Chapters 6 and 7 then showed empirically, and with specific reference to the recent development of the data processing industry, that venture capital and the issue of new equity have been among the preferred financial instruments of, respectively, new and established firms undertaking aggressive and market-oriented R&D strategies. It is therefore apparent that – as one would expect from Schumpeter's original argument – technological change in a high-tech industry such as data processing was initiated by newly established firms that raised financing of the venture capital type in order to implement their innovative activities. As the industry started to approach maturity, large firms increasingly achieved major technological advances, and funded their R&D investments by issuing new equity.

Finally, Chapter 8 showed that public policies aimed at improving the functioning of the financial system are able to affect positively the undertaking of aggressive R&D strategies in the private sector. Examined in particular was the function of microeconomic policies favourable to venture capital, of policies encouraging R&D cooperative agreements, and of R&D tax policies; that is, the principal measures intended to channel the largest possible amount of financial resources to R&D. Among the first group of policies, I stressed the importance of tax shelters for outside investors in innovating firms. In the case of R&D cooperation, I recommended instead the introduction of incentive schemes (such as financial backing for the formation of alliances among small firms). In the case of R&D tax policies affecting publicly traded companies, I suggested that in most countries corporate tax should be reduced in order to encourage internal financing and the use of equity instead of debt when external financing is sought. In this way, tax benefits do not affect the investment decisions of managers, and shareholders can be taxed only on the basis of the earnings attributable to their shares. Managers will therefore select amongst different investment strategies in order to maximise their company's expected value, whereas shareholders will take direct advantage of the company's market performance and valuation.

Although the theoretical framework of the present study belongs to the Schumpeterian tradition, I have focused more closely than Schumpeter and his followers did on the financing of the various stages of inventive activity leading up to the introduction of viable innovations. For this purpose I have deliberately adopted a multidisciplinary approach – at least with respect to the various specialisations that have emerged in the broad field of economics – to demonstrate from a theoretical, historical and empirical perspective the existence of a close interdependence between financial styles and technological styles at the macroeconomic level, and between firm financial structure and the undertaking of specific R&D activities at the microeconomic one.

Although most economists are not sympathetic to the employment of a multidisciplinary approach, I contend that multidisciplinarity in economic analysis does not necessarily entail eclectism. In this respect, one may recall that it was Schumpeter himself who emphasised that economic analysis becomes a science only when and if the three fundamental tools of theory, economic history and statistics are simultaneously employed. This is what I have modestly tried to do in this book.

Notes and References

1 The Role of Finance in Technological Change

1. For a promising attempt in this direction, in particular as regards empirical applications, see Christensen's (1992) study of the role of finance in industrial innovation, which deals with the ability of the financial system in Denmark to finance innovation.
2. See, among others, Green and Shoven (1983), Switzer (1985), Fox (1987), Craig Justice (1988), Prakke (1988), Holmström (1989), Dosi (1990), Greenwald and Jovanovic (1990).
3. For a survey of the main contributions in the evolutionary perspective, see Dosi (1988).
4. See, for example, Rosenberg (1969, 1982, 1988), Nelson and Winter (1977), Abernathy and Utterback (1978), Dosi (1982), Freeman, Clark and Soete (1982), Sahal (1985), Perez (1985) Freeman and Perez (1986), Marengo (1990), Siniscalco (1990), Andersen (1991), and others. For reviews of these contributions, see Nelson and Winter (1982, Ch. 10), Teece (1986), Clark and Juma (1987) and Dosi (1988).
5. Freeman emphasises this point in contrast with Mensch (1979), who accepted Kondratieff's contention that a bunching of radical innovations during depressions causes the subsequent upswing (see Section 5.2.1 below).
6. In Perez's (1985, p. 441) words, a new techno-economic subsystem represents 'a quantum jump in potential productivity for all or most of the economy'.
7. For a definition, see Section 4.2 below.
8. This model has some points in common with the Stiglitz and Weiss (1981) model, which considers credit rationing in the banking market in a context of imperfect information on the bank's side as regards the solving capacity of firms.
9. Spence focused on education used as a device to signal productivity in labour contracts. In his model, education not only enhances the value of human capital but it also provides information (signal) on higher ability workers. Rothschild and Stiglitz (1976) extended Spence's signalling assumptions to insurance markets.

2 The Legacy of Schumpeter, and the Limits of the Schumpeterian Approach in the Analysis of the Financing of Innovation

1. Section 2.2 of the present chapter is based on Santarelli and Pesciarelli (1990). I wish to thank Enzo Pesciarelli for kindly permitting me to reproduce part of that material.
2. For a reconstruction of the debate, see Myers (1984), Allen (1985), Modigliani (1988), Stiglitz (1988).

3. Schumpeter's microeconomic analysis of entrepreneurial behaviour stresses the fact that the entrepreneurial function manifests itself in the act of introducing a new combination of factors.

4. Although one should remember the pioneering contribution of Jeremy Bentham (see Pesciarelli, 1989; Guidi, 1990).

5. This edition, in fact, differs in some respects from the subsequent German edition of 1926 (*TED1926*) and the English edition of 1934 (*TED1934*) (which was a translation of a third German reprint of the second edition) and subsequent reprints, which are those that have usually been considered by economists. In effect economists do not usually take account of the substantive differences between the first and subsequent editions of *TED*. Elliott (1983), for example, in his otherwise very interesting article on the development of Schumpeter's theory of capitalist economic development, overlooks the existence of such differences when he writes of 'a second German edition, incorporating expository alterations but no significant substantive changes' (footnote 1).

6. In the Preface to *TED1926*, Schumpeter writes that he rewrote Chapter 6 from the beginning to 'point 1' although as a matter of fact he rewrote it from point 1 to the very end of the Chapter.

7. 'If some circumstance leads to the rise of a new entrepreneur or of a new organization, for example a trust, then, empirically speaking, necessary for this purpose is "money".... [S]uch purchasing power, if the economy is in equilibrium and if, further, there are no extraordinary factors such as the requirements of war etc., may be used only for new creations – with the exception of the case where such purchasing power is shared among the economic agents in such proportions as to cause a uniform increase in prices' (*DW*, p. 420).

8. 'Most forms of action, the collection of ideas and the habitual mode of thinking in the artistic, literary and political field can generally be traced directly back to some dominant figure whose followers have continued what he began' (*TED1911*, ch. II, p. 148).

    'Entrepreneurs ... create various far-reaching plans ... but their calculations and their intentions can never determine their action beyond a certain point in the future.... What the entrepreneur has achieved is by hypothesis something new – the typical example being the establishment of a new enterprise. If the enterprise succeeds and is profitable, then the plan has been realized. If, subsequently, the entrepreneur limits himself simply to running the enterprise incorporated in a new static situation, then he is from that moment onwards a static subject of the economy. And, to tell the truth, he may decide to continue to manage his enterprise in the same routine way indefinitely. The situation changes, however, if he remains an "entrepreneur", in the sense that we have given the term, and continues to seek to introduce new combinations. Therefore, here in the dynamic field, he can only devise plans that are limited in time by the attainment of the results that they are designed to achieve. In our way of viewing things, the process must be understood thus: a change in the original static condition is generated by the entrepreneur, for example by founding a new enterprise. However, the process is concluded when this enterprise has been established and operates successfully. Furthermore, all the conditions obtaining on the establishment of the enterprise have reacted to it and have changed. Through the establishment of the enterprise and the modification brought about thereby, a new stationary situation has been produced' (*TED1911*, pp. 427–8).

3 Technology and Finance in Economic Growth and Structural Change

1. This explains why in the academic literature the expression 'financial innovation' is in most cases used to denote the employment of a certain financial instrument within a market or an industrial sector where it was not previously in use.
2. For example, Mayer's (1990) data show that bank financing is of major importance in Italy.
3. 'Traditional' refers to those industrial sectors of continuing importance during the relevant period.

4 Asset Specificity, R&D Financing and the Signalling Properties of the Firm's Finance Structure

1. This chapter is partly based on Santarelli (1991).
2. Here 'previously' refers to the R&D project.
3. The case of the data processing industry is only briefly outlined in this chapter. It will be extensively analysed in Chapters 6 and 7.
4. Or when it is sufficient only for initial financing, and external funds are needed for subsequent stages of financing.
5. This is a typical case of involuntary default, which in the situation considered here is likely to occur if the *initial* market performance of the new product is poorer than expected.
6. The information on Genentech used in this paper is based on an interview with the company's cofounder and chairman of the board, Mr Robert Swanson, conducted on 14 January 1991, and on the company's annual reports.
7. The information on Lotus is taken from material prepared for class discussion at Harvard Business School. I wish to express my thanks to William A. Sahlman for providing the material and granting permission to use it in the present work.
8. For example, for its first product Genentech employed customised equipment made by the Swedish company Alfa-Laval. As far as human asset specificity is concerned, when starting a new project Genentech usually hires young scientists just out of post-doctoral programmes with specific training in that field.
9. In the case of Genentech and other smaller biotechnology companies located in the San Francisco Bay area, all of these locational factors seem to have played a role.
10. Initially incorporated as Micro Finance Systems.
11. That is, to a contract providing finance only over the foreseeable future.

12. That is, without guaranteeing any results.
13. In the language of venture capitalists this includes: *seed financing* (used to test a scientific or a technological concept); *start-up financing* (necessary for product development, as in the case here); *first-stage investing* (used when the prototype has been already developed and the company begins manufacturing).
14. It is worth noting, however, that the sources of the agency problem differ from those usually considered in standard principal–agent literature (Rees, 1985). In fact, in the situation that concerns us here, the agency problem is created by the risk-taking incentives implicit in the contract, whereas in standard principal–agent literature the agency problem arises because the principal has to induce a risk-averse agent to undertake a certain level of action.
15. For informational reasons, venture capitalists have at least one member on the board of directors of the firm they are funding. This is because they do not invest for short-term interest or dividends, but rather maintain a long-term capital gain orientation towards their activities.
16. Otherwise, should the principal possess the same knowledge as the agent concerning $\sigma \in \theta$, he/she will implement the R&D project directly, developing his/her sponsoring company in order to share costs and risks with other outside investors.
17. Because it is the nature of the assets that determines the features and the outcome of the contractual relationship.
18. Of course, in the case of equity joint ventures and cross holdings the reward may also take the form of rights to develop the new product further or to market it in a particular geographical area.
19. Recent extensions of this approach have focused mainly on the problems of overinvestment (Jensen, 1986; Stulz, 1990) and failure to liquidate (Harris and Raviv, 1990).
20. Even if the relationship between top managers' shareholdings and firm's performance is proved to be curvilinear.
21. The underlying assumption is again that when top executives hold a significant shareholding in the firm, they improve their performance (see Morck, Shleifer and Vishny, 1988).
22. The commercial exploitation of the new invention should in fact create extra profits for the firm.

5   The Long-Term Dynamics of Financial and Technological Styles

1. See, among others, Rostow (1975), Mensch (1979), Van Duijn (1979), Freeman, Clark and Soete (1982), Kleinknecht (1987), Solomou (1987), Tylecote (1989, 1993, 1994).
2. These are: Kondratieff (1935), De Wolff (1929), Von Ciriacy-Wantrup (1936), Schumpeter (1939), Clark (1944), Dupriez (1947; 1978), Rostow (1978), Mandel (1980), Van Duijn (1983), Bouvier (1974), Amin (1975), Kuczynski (1980).
3. A mathematical formulation of the Bieshaar and Kleinknecht model can be found in Kleinknecht (1987, pp. 20–2).

4. Baker's data have been widely used in empirical studies of long-run patterns in patent activity. See, in particular, Freeman, Clark and Soete (1982); Kleinknecht (1987).

5. The first organised system of patents was developed during the fifteenth century in Italy. Its purpose was to encourage inventive activity and to allow the state to benefit from inventions. Curiously, the first patent was granted to the famous architect Filippo Brunelleschi for the invention of a 'barge with special hoisting gear which was to be used to transport marble' (Baker, 1976, p. 8). The same concept of advantage was contained in the proclamation, issued in 1601 by Queen Elizabeth, which first introduced an organised patent system into Britain. Successive improvements were made to the British patent system as patents became increasingly numerous, the purpose being to enable more precise specification of inventors' claims. The data selected by Baker relate to a period from 1691 onwards, when the patent system was sufficiently developed to enable clear distinctions to be drawn among different patented inventions.

6. Most of the patent specifications considered by Baker were published by the Patent Office in separate leaflets within a few weeks of acceptance.

7. Baker himself cites important inventions that have never been patented, such as the miner's safety lamp and the loose-contact microphone.

8. As aptly pointed out by Rosenberg (1976, Chapter 11), the diffusion process does not entail pure replication and imitation of certain innovations. It may involve a string of further innovations introduced by an increasing number of firms as they learn the new technology and become involved in the 'swarming process'.

9. In his study of economic fluctuations, Tinbergen (1951) identified various components in the behaviour of a time series (trend, cyclical fluctuations, seasonal fluctuations, and random changes) and asserted that 'As a consequence of the irregular form of most waves, their length often cannot be stated with certainty, or at any rate it is liable to changes. Some waves are not genuine waves.... The so-called "long business-cycle waves" (Kondratiefs), which have a period of about 40 years, are, according to some authors, only more or less accidental alternations of rising and falling movements.... [W]e think that our classification of components is fundamentally sound, in particular the isolation of seasonal fluctuations, of the short random fluctuations, and of a theoretically justified trend, because separate complexes of causes can be made responsible for their presence.... So far as the distinction based on the length of the periods makes sense – which is often the case – a general method can be given to determine all the components, namely that of *moving averages*. The seasonal fluctuations and all shorter fluctuations can be eliminated' (Tinbergen, 1951, pp. 54–6).

10. Freeman, Clark and Soete (1982, ch. 3) take the *master* patents (that is, 'the first to be economically viable') in Baker's data base and plot their ten-year moving averages from 1775 to 1965. From this simple analysis it emerges that the clusters of basic inventions are not systematically related to depression phases in the Kondratieff long waves. Kleinknecht (1987, ch. 4) divides Baker's data into two categories: 'pure' process innovations (representing primarily factor saving capital equipment, such as numerically controlled machines), and product innovations (ranging from new

pharmaceutical drugs to new investment goods designed to provide new goods or services to final consumers, and from new materials to scientific instruments). The plot of the three-year and nine-year moving averages of these two series reveals that product-related patents show fluctuations that seem to fit into the scheme of Kondratieff long waves.

11. Rather inconclusive evidence is instead provided by the analysis of skewness.

12. Arkwright's patent, granted in 1769, was for 'machinery for the making of weft or yarn from cotton, flax and wool' (Baker, 1976, p. 116). It should solve most of the practical problems connected with roller drawing.

13. Robert Fogel (1964) argues that railroads played a secondary role in US economic development, although he underestimates the importance of this sector for capital formation.

14. Universal banks have significantly facilitated investment in capital-intensive industries, chemicals in particular.

15. Banca Italiana di Sconto was created in 1914, on the model of the German universal bank, mainly to meet the credit requirements of industrial firms. It was soon involved in financing the Italian weapons industry, and started a long-term partnership with the leader company in the field, Società Ansaldo. This company expanded considerably during the First World War, and in 1918 employed more than 70 000 workers in very diversified activities, ranging from engineering to mining, from building to aeronautics. The dramatic growth of Ansaldo was financed mainly by credit granted by Banca Italiana di Sconto, which by 1918 was no longer able to satisfy the financial requirements of its partner. Owing to the liquidity problems of Banca Italiana di Sconto, Ansaldo attempted, by means of a hostile takeover, to extend its control over another important bank, Banca Commerciale Italiana. When in 1920 – as a consequence of a series of financial and political setbacks – Ansaldo's attempt failed, the situation reverted to that of 1918, with the industrial group entirely dependent financially on Banca Italiana di Sconto. Things deteriorated rapidly, since Banca Italiana di Sconto could not extend its loans to Ansaldo. In December 1921 a bank panic induced customers to withdraw their deposits from Banca Italiana di Sconto, which soon went bankrupt. This enlightening story, which highlights the risk of instability entailed by cross-participation between banks and industrial firms, is analysed in depth by Sraffa (1922).

16. Houdry had in fact been able to produce petrol for internal combustion engines since 1927, although commercial exploitation of the discovery was not possible until 1936.

17. In fact a gold standard system was reintroduced in most countries during the 1920s. However this was a gold exchange standard, based on indirect convertibility with gold. With this system, monetary authorities built a double line of defence, one consisting of foreign currencies and short-term credits on foreign markets, the other represented by gold reserves.

18. The Federal Reserve System had been already established in 1913. Analogously, in most countries a central bank had been established much earlier than the period under consideration in this section. Chronologically, we have the following sequence in the creation of central banks: Sweden, 1668; England, 1694; France, 1800; Holland, 1814; Austria, 1816;

Denmark, 1818; Belgium, 1850; Germany, 1875; Japan, 1882; Italy, 1893; Switzerland, 1905. Some of these did not act as central banks in the modern sense until a considerable time after their foundation.

19. The potential impact of these new technologies has been investigated in several works (see, for example, Freeman, Clark and Soete, 1982; Pavitt, Townsend and Robson, 1986; Rothwell and Zegveld, 1985; Orsenigo, 1989), some of which deal with the role played by newly established small firms in the first commercial exploitation of the new technological style. The features of a technological style based on microelectronics and biotechnologies are investigated in depth by Tylecote (1993, 1994).

20. This index has been estimated employing a procedure analogous to that described in Section 5.2.2.

21. Note that independent firms and non-independent units belonging to this size class respectively account for 17.0 per cent and 23.4 per cent of total innovations introduced by all firms and units during the relevant period.

22. In this respect are the cases of Apple Computer Inc. and Compaq Computer Co. in the field of PCs, and those of Microsoft Co., Lotus Development Co. and Borland in that of software. See Section 7.3 below for further details.

23. It must be also noted that in some countries the process of reform in the financial framework is in this period characterised by a resurgence of 'universal banks', which have been shown in Section 5.3.3 to have represented an important financial institution during the second industrial revolution.

6 The Structure of the Data Processing Industry in the 1980s and Early 1990s

1. This contributes to explanation of the 1994 'price war' initiated by the world leader in this branch, Compaq Computer Co., and soon entered by the most important producers of personal computers. This reinforced price competition in the PC field has brought about a 25 per cent reduction in market prices.

2. At the time of writing (1994) there are signs of recovery in the US and, to a lesser extent, the European markets. However as regards Europe, 1993 can be considered the worst ever year for the DP market.

3. The slowdown in total sales of desktop computers in the period 1990–1 was accompanied by the increasing popularity and sale of notebook computers, shipments of which grew by more than 200 per cent between 1990 and 1993 (see Siegmann, 1990).

4. Most previous research on information technology focused on either computers or semiconductors. This approach has had the undoubted merit of providing a careful and detailed analysis of the inner features of each of these industries, but it has probably underestimated the importance of interindustry technology flows as sources of further advances in technology and, broadly speaking, the fact that technological innovations in the field of information technology usually have repercussions far beyond the nominal boundaries of a single sector.

5. For more than two decades, technological progress in this field has been extraordinarily rapid, but evenly distributed. According to what is known as Moore's Law (from the name of Intel's cofounder and chairman, Gordon Moore), the number of components per chip has nearly doubled every 18 months! As well as this process of downsizing, during the same period the hardware cost of processing information has dropped by about 28 per cent per year in real terms (see Flamm, 1987).

6. Under this agreement, in the short run the two companies will also develop a common hardware platform based on a version of IBM's RS/6000 chip, which will be produced by Motorola in a joint venture (see Langlois, 1992). However new talks, started in the summer of 1994, are likely to render even closer the cooperation between IBM and Apple in the coming years.

7. Of course, outsourcing is a different service to customers provided by large companies. It is expected to become a big business for companies such as IBM and Digital in the USA, and Finsiel in Europe.

8. In particular, cheap minis (priced at less than $5000) are expected to increase their market share dramatically by 1996, when they will become an ideal integration element for powerful (more than 100 Mips) PC-based client systems.

9. With IBM, NEC and Fujitsu competing against Cray to attain the technological leadership in this field.

10. For an analysis of the relationship between the computer and software industries, see Malerba, Torrisi and von Tunzelmann, 1991).

11. As stressed in Section 5.3.5, in the 1980s it was fast-growing, newly established firms that stimulated the overall process of innovation in most advanced industries.

12. With the exception of the C_{20} ratio, which decreased significantly in 1991 with respect to the previous year.

13. The figures for Australia, which are based on national estimates, sound very peculiar. However, unless statistical problems not mentioned by OECD have altered the process of data collection, they have to be taken as realistic.

14. According to the *Datamation* classification, DP revenues are defined as being derived from sales of the following: computer systems, including mainframes, minicomputers, microcomputers and personal computers, workstations, word processors, office systems and CAD/CAM systems; peripherals, including terminals, printers, plotters, disk drives, tape drives, magnetic media, and data entry devices; software, including operating systems and applications programmes; data communications equipment, including communication processors, local area networks, digital PBXs, multiplexors, modems, and facsimile machines; data services, including custom programming, systems integration, consulting, time-sharing, and remote processing; maintenance and repair; computer leasing; point-of-sale systems; and automatic teller machines. In practice the *Datamation* rankings consider those goods and services that belong to the Standard Industrial Classification (SIC) codes 3573 (electronic-computing equipment), 3574 (calculating and accounting machines), and 7372, 7374 and 7379 (computer programming and software, data processing services, and computer-related services). Among the goods belonging to the information technology *filiere*, which are explicitly excluded from this classification, the most important are therefore semiconductors and printed circuit boards.

15. However it must be pointed out that the ranking could be biased towards US companies, since *Datamation* is published in that country.
16. Although still barely visible, in China too the national DP industry is making its first steps. The booming Chinese domestic computer market is dominated by US companies, but the local maker, Legend Holdings Ltd, has a significant share of this market (40 per cent in 1994, according to company insiders).
17. In 1994 Taiwanese companies controlled one half of the world market in scanners and peripherals, nearly 30 per cent of that in network cards and terminals, and about 10 per cent of that in PCs. Contributing significantly to the success of Taiwanese DP companies has been the return to their home country of thousands of graduate engineers from American universities.

7 The Financing of Innovative Activities in the Data Processing Industry

1. In the same year, a venture capital company backed the group of scientists who created Fairchild Semiconductor, out of which mushroomed the entire San Francisco Bay area semiconductor complex now called Silicon Valley (see Rind, 1985). As far as the US data processing industry is concerned, venture capital has played a crucial role in the formation and nurturing of companies such as Apple Computer, Compaq Computer, Conner Peripherals, Cray Research, Data General, Digital Equipment, Lotus Development, Microsoft, Seagate Technology, Sun Microsystems, Tandem Computer and Xerox.
2. The milestone of US venture capital was probably when Sherman Fairchild provided the funds needed to put together three small office equipment companies, thereby giving rise to IBM in 1911. However the origins of the modern era of venture capital are generally traced back to the Rockefeller family, which began its successful investment activities by financing Eastern Airlines in 1938.
3. Information on the early development of the US venture capital industry in taken mainly from Soussou (1985).
4. The era of publicly traded companies in the venture capital market came to an end in 1975, when only 16 SBICs survived.
5. The Commodore PET was introduced in 1977 but it never rivalled Apple in terms of sales, partly because of the company's slowness in shipping reliable products. Commodore soon chose a strategy different from Apple's, and focused primarily on the low end of the market, represented by home computers. The other entrant in 1977 was Tandy Computer, which used its chain of electronics stores (Radio Shack) to sell a microcomputer called TRS-80 Model I. Radio Shack stores had shipped 3000 units of this machine by the end of 1977 (see Langlois, 1992).
6. In the early 1980s some of the largest suppliers of computer devices, such as IBM and Control Data, were almost completely vertically integrated. However the diffusion of the 5.25-inch disk drive initiated a process of vertical disintegration, with a number of specialised suppliers of this kind of peripheral: Ampex, Computer Memories, International Memories, Memorex, Micro Peripherals, Miniscribe, Mitsubishi Electric Co., New

World Computer, Nippon Peripherals, Olivetti OPE, Rodime, Rotating Memory Systems, Seagate Technology, Shugart Associates, Siemens, SLI Industries, Tandon, Texas Instruments, Vertex Peripherals.

7. Information on this company is derived mainly from case studies prepared for class discussion at Harvard Business School, and from trade publications. I wish to express my thanks to William Sahlman for allowing me to use this material.

8. The sample is composed of eleven US companies: Data General, Digital Equipment Corporation, Hewlett-Packard, Honeywell, IBM, NCR, National Semiconductors, Unisys, Wang, Xerox and Zenith; one Canadian: Northern Telecom; three European: Olivetti (Italy), Nokia (Finland) and Plessey (UK); three Japanese: Hitachi, NEC and Toshiba.

9. In this respect, see the reconstruction of this argument presented in Section 2.3 above.

10. MBO funds are similar in structure to venture capital funds, as confirmed by the fact that many venture capital firms invest in leveraged buy-outs.

11. Unix International and OSF represent further steps in the struggle between AT&T and IBM, each declaring its version of UNIX to be the standard.

8 Public Policies towards the Financing of Innovative Activities

1. Which are Canada, France, Germany, Italy, Japan, the UK and the USA.

2. Since firms in the USA are taxed on their worldwide income, these credits are of course allowed only for taxes paid to foreign governments.

3. The fact that Japanese banks can hold shares in a firm reduces their lending risks. In effect, being creditors and stockholders at the same time, they are indifferent to making the firm riskier than anticipated by creditors and to transferring wealth from them to shareholders (Holder and Tschoegl, 1990).

4. In this market orientation of policy towards innovation in Japan, government procurement represented a large share (about 50 per cent) of total demand for computers in the 1960s, when Japanese DP devices were still less advanced than the corresponding foreign products (see Flamm, 1987).

5. Mansfield (1986) has shown that, when used without discriminating between one market and another, tax-based incentives to R&D have not been very effective. However this finding does not hold for those countries that are 'lagging behind' in terms of R&D commitment and need to reach an optimal level of innovative activities

Bibliography

The references comprise works directly quoted in the text and tables and do not acknowledge contributions of ideas and evidence from various other sources. In particular, extensive use has been made of the following periodicals: *Business Week, Datamation, Electronics, Financial Times, Fortune, International Herald Tribune, Il Sole-24Ore* and *Mondo Economico*.

Abernathy, W. J., K. B. Clark and A. M. Kantrow (1983) *Industrial Renaissance. Producing a Competitive Future for America* (New York: Basic Books).

Abernathy, W. J. and J. M. Utterback (1978) 'Patterns of Industrial Innovation', *Technology Review*, vol. 7, no. 80, pp. 40–7.

Abramovitz, M. (ed.) (1955) *Capital Formation and Economic Growth* (Princeton, NJ: Princeton University Press for NBER).

Aghion, P. and P. Bolton (1992) 'An Incomplete Contracts Approach to Financial Contracting', *Review of Economic Studies*, vol. 59, no. 3, pp. 473–94.

Aglietta, M. (1976) *Régulation et crises du capitalisme. L'exemple des Etats-Unis* (Paris: Calmann-Lévy).

Akerlof, G. A. (1970) 'The Markets for "Lemons": Qualitative Uncertainty and the Market Mechanism', *Quarterly Journal of Economics*, vol. 84, no. 4, pp. 488–500.

Allen, D. E. and H. Mizuno (1989) 'The Determinants of Corporate Capital Structure: Japanese Evidence', *Applied Economics*, vol. 21, no. 5, pp. 569–85.

Allen, F. (1985) 'Capital Structure and Imperfect Competition in Product Markets', Working Paper, The Wharton School, University of Pennsylvania.

Amendola, M. (ed.) (1990) *Innovazione e progresso tecnico* (Bologna: il Mulino).

Amin, S. (1975) *Une crise structurelle. La crise de l'impérialisme* (Paris: Editions de Minuit).

Andersen, E. S. (1991) 'Techno-economic Paradigms as Typical Interfaces Between Producers and Users', *Journal of Evolutionary Economics*, vol. 1, no. 2, pp. 119–44.

Aoki, M. (1985) 'The Macroeconomic Background for High-Tech Industrialization in Japan', in Landau and Rosenberg (eds), pp. 569–581.

Aoki, M. (1988) *Information, Incentives, and Bargaining in the Japanese Economy* (New York: Cambridge University Press).

Aoki, M. (1990) 'Toward an Economic Model of the Japanese Firm', *Journal of Economic Literature*, vol. 28, no. 1, pp. 1–27.

Appleton, E. L. (1991) 'Ouch! Europe's Biggest Suppliers Feel the Pinch', *Datamation*, vol. 37, no. 13, pp. 60–4.

Arrow, K. J. (1962) 'Economic Welfare and the Allocation of Resources for Inventions', in Nelson (ed.), pp. 609–25.

Arrow, K. J. (1985) 'The Economics of Agency', in Pratt and Zeckhauser (eds), pp. 37–51.

Arthur, W. B. (1983) 'On Competing Technologies and Historical Small Events: The Dynamics of Choice Under Uncertainty', paper presented at the Technological Innovation Program Workshop, Department of Economics, Stanford University.

Arthur, W. B. (1990) 'Silicon Valley Locational Clusters: When Do Increasing Returns Imply Monopoly?', *Mathematical Social Sciences*, vol. 19, no. 3, pp. 235–51.

Arthur, W. B. (1989) 'Competing Technologies, Increasing Returns, and Lock-in by Historical Events', *Economic Journal*, vol. 99, no. 1, pp. 116–31.

Assinform (various years) *Rapporto sulla situazione dell'informatica in Italia* (Milan: Promobit).

Auerbach, A. J. (ed.) (1988) *Corporate Takeovers: Causes and Consequences* (Chicago: University of Chicago Press).

Baker, J. (1976) *New & Improved. Inventors and Inventions That Have Changed the World* (London: British Museum Publications Limited).

Berle, A. and G. Means (1932) *The Modern Corporation and Private Property* (New York: Macmillan).

Bewley, T. (ed.) (1987) *Advances in Economic Theory, Fifth World Congress* (Cambridge: Cambridge University Press).

Bieshaar, H. and A. Kleinknecht (1983) 'Kondratieff Long Waves in Aggregate Output? An Econometric Test', *Koniunkturpolitik*, vol. 30, no. 5.

Bouvier, J. (1974) 'Capital bancaire, capital industriel et capital financier dans la croissance du XIXe siècle', *La Pensée*, no. 178.

Boyer, R. and J. Mistral (1978) *Accumulation, inflation, crises* (Paris: Presses Universitaires de France).

Brown, M. (1966) *On the Theory and Measurement of Technological Change* (Cambridge: Cambridge University Press).

Burgelman, R. and R. Rosenbloom (eds) (1989) *Technology, Competition and Organization Theory* (Greenwich, CT: JAI Press).

Cameron, R., (ed.) (1967) *Banking in the Early Stages of Industrialization* (Oxford: Oxford University Press).

Carlsson, B. (ed.) (1989) *Industrial Dynamics* (Boston: Kluwer Academic Publishers).

Chandler, A. D. Jr. (1965) *The Railroads: The Nation's First Big Business* (New York: Doubleday).

Chandler, A. D. Jr. (1966) *Strategy and Structure* (New York: Doubleday).

Christensen, J. L. (1992) *The Role of Finance in Industrial Innovation* (Aalborg: Aalborg University Press).

Clark, C. (1944) *The Economics of the 1960s* (London: Macmillan).

Clark, C. (1984) 'Is There a Long Cycle?', *BNL – Quarterly Review*, vol. 37, no. 150, pp. 307–20.

Clark, K.B. and R. Hayes (eds) (1985) *The Uneasy Alliance. Managing the Productivity–Technology Dilemma* (Cambridge, Mass.: Harvard Business School Press).

Clark, N. and C. Juma (1987) *Long-run Economics: An Evolutionary Approach to Economic Growth* (London: Pinter).

Commons, J. (1924) *Legal Foundations of Capitalism* (Madison: University of Wisconsin Press).

Commons, J. (1934) *Institutional Economics – Its Place in Political Economy* (Madison: University of Wisconsin Press).

Craig Justice, S. (1988) 'The Financial Linkages Between the Development and Acquisition of Technology', *Journal of Economic Issues*, vol. 22, no. 2, pp. 355–62.

Crouzet, F. (ed.) (1972) *Capital Formation in the Industrial Revolution* (London: Methuen).

Dahmèn, E. (1984) 'The Entrepreneur and Economic Transformation', *Journal of Economic Behavior and Organization*, vol. 5, no. 1, pp. 25–34.

Dasgupta, P. and J. Stiglitz (1980) 'Uncertainty, Industrial Structure, and the Speed of R&D', *Bell Journal of Economics*, vol. 11, no. 1, pp. 1–28.

David, P. (1986) 'Understanding the Economics of QWERTY: The Necessity of History', in Parker (ed.).

David, P. (1988) 'Path-dependence: Putting the Past into the Future of Economics', Stanford University, Institute for Mathematical Studies in the Social Sciences, technical report no. 533.

David, P. and E. Steinmueller (1990) 'The ISDN Bandwagon is Coming, But Who Will Be There to Climb Aboard? Quandaries in the Economics of Data Communication Networks', *Economics of Innovation and New Technology*, vol. 1, no. 1, pp. 43–62.

Dean, B. V. and J. J. Giglierano (1990) 'Multistage Financing of Technical Start-up Companies in Silicon Valley'; *Journal of Business Venturing*, vol. 5, no. 6, pp. 375–89.

De Cecco, M. (1974) *Money and Empire. The International Gold Standard, 1890–1914* (Oxford: Basil Blackwell).

DEFTA (1990) 'VC in US–Japan Market', *Microtimes*, 6 August.

De Roover, R. (1954) 'New Interpretations of the History of Banking', *Cahiers d'histoire mondiale*, vol. 2.

Devine, W. D. (1983) 'From Shafts to Wires: Historical Perspectives on Electrification', *Journal of Economic History*, vol. 43, no. 2, pp. 347–72.

De Wolff, S. (1929) *Het economische getij* (Amsterdam: J. Emmering).

Di Matteo, M., R. M. Goodwin and A. Vercelli (eds) (1989) *Technological and Social Factors in Long Term Fluctuations* (Berlin: Springer-Verlag).

Dollar, D. and J. Frieden (1989) 'The Political Economy of Financial Deregulation in the United States and Japan', in Luciani (ed.), pp. 73–102.

Donaldson, G. (1961) *Corporate Debt Capacity* (Boston, Mass.: Harvard Graduate School of Business).

Dosi, G. (1982) 'Technological Paradigms and Technological Trajectories: a Suggested Interpretation of the Determinants and Directions of Technical Change', *Research Policy*, vol. 11, no. 2, pp. 147–62.

Dosi, G. (1984) 'Technological Paradigms and Technological Trajectories: The Determinants and Directions of Technical Change and the Transformation of the Economy', in Freeman (ed.), pp. 78–101.

Dosi, G., C. Freeman, R. Nelson, L. Soete and G. Silverberg (eds) (1988) *Technical Change and Economic Theory* (London: Frances Pinter).

Dosi, G. (1988) 'Sources, Procedures and Microeconomic Effects of Innovation', *Journal of Economic Literature*, vol. 26, no. 3, pp. 1120–71.

Dosi, G. (1990) 'Finance, Innovation and Industrial Change', *Journal of Economic Behavior and Organization*, vol. 13, no. 3, pp. 299–319.

Dupriez, L. H. (1947) *Des mouvements économiques généraux* (Louvain: Institut de Recherches Economiques et Sociales de l'Université de Louvain).

Dupriez, L. H. (1978) 'A Downturn in the Long Wave?', *Banca Nazionale del Lavoro – Quarterly Review*, vol. 31, no. 126.

EC Commission (1985) *Fourteenth Report on Competition Policy* (Brussels: EC).

EC Commission (1986) *Incentives for Industrial Research, Development, and Innovation* (London: Kogan Page).

Economist, The (1991) 'Big Blue Apple', 5 October, pp. 97–98.

Elliott, J. E. (1983) 'Schumpeter and the Theory of Capitalist Economic Development', *Journal of Economic Behavior and Organization*, vol. 4, no. 4, pp. 277–308.

Elton, E. J. and M. J. Gruber (eds) (1990) *Japanese Capital Markets* (New York: Harper & Row).

Enos, J. (1962) 'Invention and Innovation in the Petroleum Refining Industry', in Nelson (ed.).

Etzioni, A. (1987) 'Entrepreneurship, Adaptation, and Legitimation', *Journal of Economic Behavior and Organization*, vol. 8, no. 2, pp. 175–89.

Evan, W. and P. Oik (1990) 'R&D Consortia: A New US Organizational Form', *Sloan Management Review*, pp. 37–46.

EVCA (European Venture Capital Association) (various years) *Venture Capital Year Book*, Zaventem (Belgium).

Fama, E. F. and M. H. Miller (1972) *The Theory of Finance* (New York: Holt Rinehart and Winston).

Flamm, K. (1987) *Targeting the Computer. Government Support and International Competition* (Washington, DC: The Brookings Institution).

Fogel, R. W. (1964) *Railroads and American Economic Growth: Essays in Econometric History* (Baltimore: The Johns Hopkins University Press).

Fox, M. B. (1987) *Finance and Industrial Performance in a Dynamic Economy* (New York: Columbia University Press).

Francis, B. (1989) 'The Destkop Dimension' *Datamation*, 1 November, pp. 28–40.

Freeman, C. (ed.) (1984) *Long Waves in World Economy* (London: Frances Pinter).

Freeman, C. (1987) *Technology Policy and Economic Performance* (London: Frances Pinter).

Freeman, C., J. Clark and L. Soete (1982) *Unemployment and Technical Innovation* (London: Frances Pinter).

Freeman, C. and C. Perez (1986) 'The Diffusion of Technical Innovation and Changes of Techno-Economic Paradigm', paper presented at the Conference on Innovation Diffusion, held in Venice, DAEST, March.

Freeman, C. *et al.* (eds) (1991) *Technology and the Future of Europe* (London: Frances Pinter).

Friedman, B. M. (ed.) (1985) *Corporate Capital Structures in the United States* (Chicago: University of Chicago Press).

Friedman, J. W. (1986) *Game Theory With Applications to Economics* (Oxford: Oxford University Press).

Friedman, M. and A. Schwartz (1963) *A Monetary History of the United States, 1867–1960* (Princeton, NJ: Princeton University Press for NBER).

Geroski, P. (1993) 'Antitrust Policy Towards Co-operative R&D Ventures', *Oxford Review of Economic Policy*, vol. 9, no. 2, pp. 58–71.

Gibbons, R. and K. J. Murphy (1990) 'Relative Performance Evaluation for Chief Executive Officers', *Industrial and Labor Relations Review*, vol. 43, no. 3, pp. 30/S–51/S.

Giersch, H. (ed.) (1982) *Emerging Technologies* (Tubingen: J. C. B. Mohr).

Goldsmith, R. W. (1955) 'Financial Structure and Economic Growth in Advanced Countries', in Abramovitz (ed.)

Goldsmith, R. W. (1969) *Financial Structure and Development* (New Haven, CT: Yale University Press).

Goldsmith, R. W. (1971) 'The Development of Financial Institutions During the Post-War Period', *BNL – Quarterly Review*, vol. 22.

Gomulka, S. (1990) *The Theory of Technological Change and Economic Growth* (London and New York: Routledge).

Goodwin, R. M. (1986) 'Towards a Theory of Long Waves', paper presented at the International Workshop on Technological and Social Factors in Long Term Fluctuations, Siena (Italy), 16–18 December, now in Di Matteo *et al.* (1989), pp. 1–15.

Grasso, M. (1989) 'Merger & Acquisition nel settore Information Technology nel 1988', mimeo, Olivetti Systems & Network.

Green, J. R. and J. B. Shoven (1983) 'The Effects of Financing Opportunities and Bankruptcy on Entrepreneurial Risk-Bearing', in Ronen (ed.).

Greenwald, B. and B. Jovanovic (1990) 'Financial Development, Growth and the Distribution of Income', *Journal of Political Economy*, vol. 98, no. 5, part 1.

Greenwald, B. and J. E. Stiglitz (1992) 'Information, Finance, and Markets: The Architecture of Allocative Mechanisms', *Industrial and Corporate Change*, vol. 1, no. 1, pp. 37–63.

Guidi, M. E. L. (1990) 'Shall the Blind Lead Those Who Can See? Bentham's Theory of Political Economy', in Moggridge (ed.), vol. 3, pp. 10–28.

Gurley, J. G. and E. S. Shaw (1957) 'The Growth of Debt and Money in the United States: A Suggested Interpretation', *Review of Economics and Statistics*, vol. 39.

Gurley, J. G. and E. S. Shaw (1960) *Money in a Theory of Finance* (Washington, DC: The Brookings Institution).

Hall, B. (1988) 'The Effects of Takeover Activity on Corporate Research and Development', in Auerbach (ed.), pp. 69–96.

Hall, B. (1993) 'R&D Tax Policy During the 1980s: Success or Failure?', University of California at Berkeley, Department of Economics, working paper no. 93–208.

Hansmann, H. and R. Kraakman (1992) 'Hands-Tying Contracts: Book Publishing, Venture Capital Financing, and Secured Debt', *Journal of Law, Economics, & Organization*, vol. 8, no. 3, pp. 628–55.

Harris, M. and A. Raviv (1979) 'Optimal Incentive Contracts with Imperfect Information', *Journal of Economic Theory*, vol. 20, no. 1, pp. 231–59.

Harris, M. and A. Raviv (1990) 'Capital Structure and the Informational Role of Debt', *Journal of Finance*, vol. 45, no. 2, pp. 321–49.

Hart, O. (1988) 'Capital Structure as a Control Mechanism in Corporations', *Canadian Journal of Economics*, vol. 21, no. 3, pp. 467–76.

Hart, O. and B. Holmström (1987) 'The Theory of Contracts', in Bewley (ed.), pp. 71–155.

Hawke, G. R. (1970) *Railways and Economic Growth in England and Wales, 1840–1870* (Oxford: Clarendon Press).

Heertje, A. (ed.) (1988) *Innovation, Technology, and Finance* (Oxford: Basil Blackwell).

Hodgson, G. (1986) 'Behind Methodological Individualism', *Cambridge Journal of Economics*, vol. 10, no. 3, pp. 211–24.

Holder, J. E. and A. E. Tschoegl (1990) 'Some Aspects of Japanese Corporate Finance', in Elton and Gruber (eds), pp. 57–80.

Holmström, B. (1979) 'Moral Hazard and Observability', *Bell Journal of Economics*, vol. 10, no. 1, pp. 74–91.

Holmström, B. (1989) 'Agency Costs and Innovation', *Journal of Economic Behavior and Organization*, vol. 12, no. 3, pp. 305–27.

Hoshi, T., A. Kashyap and D. Scharfstein (1990) 'The Role of Banks in Reducing the Costs of Financial Distress in Japan', NBER Working Paper no. 3435.

Hubbard, R. G. (ed.) (1990) *Asymmetric Information, Corporate Finance, and Investment* (Chicago, ILL: The University of Chicago Press for NBER).

IG (1983) 'Business Outlook: Hard Disk Drives', *High Technology*, January, p. 53.

Jensen, M. C. (1986) 'Agency Costs of Free Cash Flow, Corporate Finance and Takeovers', *American Economic Review*, vol. 76, no. 2, pp. 323–39.

Jensen, M. C. and W. H. Meckling (1976) 'Theory of the Firm: Managerial Behaviour, Agency Costs, and Ownership Structure', *Journal of Financial Economics*, vol. 3, no. 4, pp. 305–60.

Jewkes, J., D. Sawers and R. Stillerman (1962) *The Sources of Invention* (London: Macmillan).

Jorde, T. M. and D. J. Teece (eds) (1992) *Antitrust, Innovations and Competitiveness* (Oxford: Oxford University Press).

Judge, G., W. Griffiths, C. Hill and T. Lee (1980) *The Theory and Practice of Econometrics* (New York: John Wiley & Sons).

Kalecki, M. (1937) 'The Principle of Increasing Risk', *Economica*, vol. 4, no. 4, pp. 440–7.

Katz, B. G. and A. Phillips (1981) 'Government, Technological Opportunities and the Emergence of the Computer Industry', in Giersch (ed.), pp. 419–66.

Kaufman, H. (1986) *Interest Rates, the Market, and the New Financial World* (New York: Times Books).

Keynes, J. M. (1930) *A Treatise on Money* (London: Macmillan), 2 vols.

Keynes, J. M. (1936) *The General Theory of Employment, Interest and Money* (London: Macmillan).

Kindleberger, C. P. (1975) 'Germany's Overtaking of England, 1806–1914', *Weltwirtschaftliches Archiv*, vol. 111, nos 2 and 3.

Kindleberger, C. P. (1984) *A Financial History of Western Europe* (London: George Allen & Unwin).

Kleinknecht, A. (1987) *Innovation Patterns in Crisis and Prosperity. Schumpeter's Long Cycle Reconsidered* (London and New York: Macmillan and St. Martin's Press).

Kondratieff, N. (1935) 'The Long Waves in Economic Life', *Review of Economics and Statistics*, vol. 17, no. 1, pp. 105–15.

Kreps, D. M. (1990) *A Course in Microeconomic Theory* (Hemel Hempstead: Harvester Wheatsheaf).

Kuczynski, T. (1978) 'Spectral Analysis and Cluster Analysis as Mathematical Methods for the Periodization of Historical Processes: A Comparison of Results Based on Data about the Development of Production and Innovation in the History of Capitalism. Kondratieff Cycles: Appearance or Reality?', paper presented at the Seventh International Economic History Congress, Edinburgh.

Kuczynski, T. (1980) 'Have There Been Differences Between the Growth Rates in Different Periods of the Development of the Capitalist World Economy Since 1850? An Application of Cluster Analysis in Time Series Analysis', *Historisch-sozialwissenschaftliche Forschungen*, vol. 6.

Kuh, E. and J. R. Meyer (1959) *The Investment Decision* (Cambridge, Mass.: Harvard University Press).

Kuhn, T. (1963) *The Structure of Scientific Revolutions* (Chicago, ILL: University of Chicago Press).

Kupfer, A. (1990) 'American's Fastest-Growing Company', *Fortune*, 13 August, pp. 48–54.

Kuznets, S. (1971) *Economic Growth of Nations* (Cambridge, Mass.: Harvard University Press).

Lakatos, I. (1970) 'Falsification and the Methodology of Scientific Research Programmes', in Lakatos and Musgrave (eds).

Lakatos, I. and A. Musgrave (eds) (1970) *Criticism and the Growth of Knowledge* (Cambridge: Cambridge University Press).

Landau, R. and N. Rosenberg (eds) (1985) *The Positive Sum Strategy* (Washington, DC: National Academy Press).

Langlois, R. N. (1992) 'External Economies and Economic Progress: The Case of the Microcomputer Industry', *Business History Review*, vol. 66, no. 1, pp. 1–50.

Lawriwsky, M. L. (1984) *Corporate Structure and Performance* (London: Croom Helm).

Leland, H. E. and D. H. Pyle (1977) 'Information Asymmetries, Financial Structure, and Financial Intermediaries', *Journal Finance*, vol. 32, no. 6, pp. 1167–89.

Levin, R. C., A. K. Klevorick, R. R. Nelson and S. G. Winter (1987) 'Appropriating the Returns from Industrial Research and Development', *Brookings Papers on Economic Activity*, no. 15.

Levine, S. N. (ed.) (1985) *Investing in Venture Capital and Buyouts* (Homewood, ILL: Dow Jones-Irwin).

Lewellen, W. G. (1969) 'Management and Ownership in the Large Firm', *Journal of Finance*, vol. 24, no. 2, pp. 299–322.

Link, A. (1988) 'Acquisitions as Sources of Technological Innovation', *Mergers & Acquisitions*, November/December, pp. 36–9.

Link, A. N. and G. Tassey (eds) (1989) *Cooperative Research and Development: The Industry–University–Government Relationship* (Dordrecht: Kluwer Academic Publishers).

Luciani, G. (ed.) (1989) *La finanza americana tra euforia e crisi* (Milan: Fondazione Adriano Olivetti).

Mc Grane, R. C. (1924) *The Panic of 1837* (Chicago, ILL: University of Chicago Press).

MacMillan, I. C., D. M. Kulow and R. Khoylian (1990) 'Venture Capitalists' Involvement in Their Investments: Extent and Performance', *Journal of Business Venturing*, vol. 4, no. 1, pp. 27–47.

Maddison, A. (1977) 'Phases of Capitalist Development', *BNL – Quarterly Review*, vol. 30, no. 121, pp. 103–37.

Malerba, F., S. Torrisi and G. N. von Tunzelmann (1991) 'Electronic Computers', in Freeman *et. al.* (eds).

Mandel, E. (1980) *Long Waves of Capitalist Development. The Marxist Explanation* (Cambridge University Press).

Mansfield, E. (1986) 'The R&D Tax Credit and Other Technology Policy Issues', *American Economic Review*, vol. 76, no. 2, pp. 190–4.

Marengo, L. (1990) 'Coordination and Organizational Learning in the Firm', *Journal of Evolutionary Economics*, vol. 2, no. 44, pp. 313–26.

Martin, S. (1993) 'Public Policies Toward Cooperation in Research and Development: the European Community, Japan, the United States' (Florence: European University Institute, May).

Mayer, C. (1988) 'New Issues in Corporate Finance', *European Economic Review*, vol. 32, no. 5, pp. 1167–89.

Mayer, C. (1990) 'Financial Systems, Corporate Finance, and Economic Development', in Hubbard (ed.), pp. 307–32.

Mensch, G. (1979) *Stalemate in Technology: Innovations Overcome the Depression* (New York: Ballinger), first German edition (1975), *das technologische Patt. Innovationen uberwinden die Depression* (Frankfurt: Umschau).

Meyer, J. R. and J. M. Gustafson (eds) (1988) *The U.S. Business Corporation. An Institution in Transition* (Cambridge, Mass: Ballinger).

Minsky, H. P. (1982) *Can 'It' Happen Again? Essays on Instability and Finance* (New York: M. E. Sharpe).

Mitchell, B. R. (1964) 'The Coming of the Railway and United Kingdom Economic Growth', *Journal of Economic History*, vol. 24, no. 3, pp. 315–36.

Moad, J. (1990) 'The Challenge Is Applications', *Datamation*, 15 March, pp. 25–6.

Modigliani, F. (1988) 'MM – Past, Present, and Future', *Journal of Economic Perspectives*, vol. 2, no. 2, pp. 149–58.

Modigliani, F. and M. H. Miller (1958) 'The Cost of Capital, Corporation Finance, and the Theory of Investment', *American Economic Review*, vol. 48, no. 2, pp. 261–97.

Moggridge, D. E. (ed.) (1990) *Perspectives on the History of Economic Thought* (Brookfield: Edward Elgar), 3 vols.

Morck, R., A. Shleifer and R. Vishny (1988) 'Management Ownership and Market Valuation: An Empirical Analysis', *Journal of Financial Economics*, vol. 15, no. 3, pp. 293–315.

Mowery, D. (1983) 'Industrial Research and Firm Size, Survival, and Growth in American Manufacturing, 1921–1946: An Assessment', *Journal of Economic History*, vol. 43, no. 4, pp. 953–80.

Myers, S. C. (1977) 'Determinants of Corporate Borrowing', *Journal of Financial Economics*, vol. 5, no. 2, pp. 147–75.

Myers, S. C. (1984) 'The Capital Structure Puzzle', *Journal of Finance*, vol. 39, no. 3, pp. 575–92.

Myers, S. C. (1985) 'Comments on Investment Patterns and Financial Leverage', in Friedman (ed.), pp. 348–51.

Myers, S. C. and N. S. Majluf (1984) 'Corporate Financing and Investment Decisions When Firms Have Information That Investors Do Not Have', *Journal of Financial Economics*, vol. 13, no. 2, pp. 187–221.

Nelson, R. R. (1962) 'The Link Between Science and Invention: The Case of the Transistor', in id. (ed.)

Nelson, R. R. (ed.) (1962) *The Rate and Direction of Inventive Activity* (Princeton, NJ: Princeton University Press for NBER).

Nelson, R. R. (1991) 'The Role of Firm Differences in an Evolutionary Theory of Technical Advance', *Science and Public Policy*, vol. 18, no. 6, pp. 347–52.

Nelson, R. R. and S. G. Winter (1977) 'In Search of Useful Theory of Innovation', *Research Policy*, vol. 5, no. 1, pp. 36–76.

Nelson, R. R. and S. G. Winter (1982) *An Evolutionary Theory of Economic Change* (Cambridge, Mass.: The Belknap Press of Harvard University Press).

Oakey, R. P. (1984) 'Finance and Innovation in British Small Independent Firms', *Omega*, vol. 12, no. 2, pp. 113–24.

OECD (1989) *Biotechnology. Economic and Wider Impacts* (Paris: OECD).

OECD (1991) *Basic Science and Technology Statistics* (Paris: OECD).

Okimoto, D. I. (1985) 'The Japanese Challenge in High Technology', in Landau and Rosenberg (eds), pp. 541–67.

Orsenigo, L. (1989) *The Emergence of Biotechnology* (London: Frances Pinter).

Parker, W. N. (ed.) (1986) *Economic History and the Modern Economist* (Oxford: Basil Blackwell).

Pavitt, K., J. Townsend and M. Robson (1986) *Sectoral Patterns of Production and Use of Innovation in the UK: 1945–1983*, mimeo, Science Policy Research Unit, University of Sussex.

Peck, M. J. (1988) 'The Large Japanese Corporation', in Meyer and Gustafson (eds), pp. 21–42.

Perez, C. (1983) 'Structural Change and Assimilation of New Technologies in the Economic and Social Systems', *Futures*, vol. 15, no. 5, pp. 357–75.

Perez, C. (1985) 'Microelectronics, Long Waves, and World Structural Change', *World Development*, vol. 13, no. 3, pp. 441–63.

Pesciarelli, E. (1989) 'Smith, Bentham, and the Development of Contrasting Ideas on Entrepreneurship', *History of Political Economy*, vol. 21, no. 3, pp. 521–536.

Piergiovanni, R. and E. Santarelli (1995) 'The Determinants of Firm Start-up and Entry in Italian Producer Services', *Small Business Economics*, vol. 7, no. 3, pp. 221–30.

Piore, M. and C. Sabel (1984) *The Second Industrial Divide. Possibilities for Prosperity* (New York: Basic Books).

Pisano, G. P. and D. J. Teece (1989) 'Collaborative Arrangements and Global Technology Strategy', in Burgelman and Rosenbloom (eds).

Pitelis, C. (1987) *Corporate Capital. Control, Ownership, Saving and Crisis* (Cambridge: Cambridge University Press).

Podolski, T. M. (1986) *Financial Innovation and the Money Supply* (Oxford: Basil Blackwell).

Pollard, S. (1964) 'Fixed Capital in the Industrial Revolution', *Journal of Economic History*, vol. 24, no. 3, pp. 299–314.

Prakke, F. (1988) 'The Financing of Technical Innovation', in Heertje (ed.), pp. 71–100.

Pratt, J. W. and R. J. Zeckhauser (eds) (1985) *Principals and Agents: The Structure of Business* (Boston, Mass.: Harvard Business School Press).

Rees, M. (1985) 'Theory of Principal and Agent – Part I', *Bulletin of Economic Research*, vol. 37, no. 1, pp. 3–25.

Riley, J. G. (1975) 'Competitive Signalling', *Journal of Economic Theory*, vol. 10, no. 2, pp. 174–86.

Rind, K. W. (1985) 'Overview of Venture Capital', in Levine (ed.), pp. 1–9.

Robinson, E. A. G. (ed.) (1966) *Problems in Economic Development* (London: Macmillan).

Robson, M. and J. Townsend (1984) 'Trends and Characteristics of Significant Innovations and their Innovators in the UK since 1945', mimeo, SPRU, University of Sussex, August.

Rogers, E. M. and J. K. Larsen (1984) *Silicon Valley Fever* (New York: Basic Books).

Ronen, J. (ed.) (1983) *Entrepreneurship* (Lexington, Mass.: D. C. Heath).

Roosa, R. V. (1951) 'Interest Rates and Central Bank', in *Money, Trade and Economic Growth. Essays in Honour of John Henry Williams* (New York: Macmillan).

Rosenberg, N. (1963) 'Technological Change in the Machine-Tool Industry 1840–1910', *Journal of Economic History*, vol. 23, no. 3, pp. 414–43.

Rosenberg, N. (1969) 'The Direction of Technological Change: Inducement Mechanisms and Focusing Devices', *Economic Development and Cultural Change*, vol. 18, no. 1, pp. 1–24.

Rosenberg, N. (1972) *Technology and American Economic Growth* (Armonk, NY: M. E. Sharpe).

Rosenberg, N. (1975) 'Problems in the Economist's Conceptualization of Technological Innovation', *History of Political Economy*, vol. 7, no. 4, pp. 456–81.

Rosenberg, N. (1976) *Perspectives in Technology* (Cambridge: Cambridge University Press).

Rosenberg, N. (1982) *Inside the Black Box: Technology and Economics* (Cambridge: Cambridge University Press).

Rosenberg, N. (1985) 'The Commercial Exploitation of Science by American Industry', in Clark and Hayes (eds), pp. 1–26.

Rosenberg, N. (1988) 'Qualitative Aspects of Technological Change: Some Historical Perspectives', unpublished manuscript, Department of Economics, Stanford University.

Ross, S. A. (1977) 'The Determination of Financial Structure: The Incentive-Signalling Approach', *Bell Journal of Economics*, vol. 8, no. 1, pp. 23–40.

Rostow, W. W. (1975) 'Kondratieff, Schumpeter, Kuznets: Trend Periods Revisited', *Journal of Economic History*, vol. 35

Rostow, W. W. (1978) *The World Economy. History and Prospect* (London: Macmillan).

Rothschild, M. and J. E. Stiglitz (1976) 'Equilibrium in Competitive Insurance Markets: an Essay on the Economics of Imperfect Information', *Quarterly Journal of Economics*, vol. 90, no. 4, pp. 629–49.

Rothwell, R. (1985) 'Venture Finance, Small Firms and Public Policy in the UK', *Research Policy*, vol. 14, no. 5, pp. 253–65.

Rothwell, R. and W. Zegveld (1985) *Reindustrialization and Technology* (London: Longman).

Rybczynski, T. M. (1974) 'Business Finance in the EEC, USA and Japan', *The Three Banks Review*, no. 103, pp. 58–72.

Sahal, D. (1985) 'Technological Guideposts and Innovation Avenues', *Research Policy*, vol. 14, no. 1, pp. 61–82.

Sahlman, W. A. and H. H. Stevenson (1985) 'Capital Market Myopia', *Journal of Business Venturing*, vol. 1, no. 1.

Sahlman, W. W. (1990) 'Insights from the American Venture Capital Organization', mimeo, Graduate School of Business Administration, Harvard University.

Santarelli, E. (1985) 'A proposito di una nuova raccolta di scritti schumpeteriani', *Studi Economici*, no. 26, pp. 137–47.

Santarelli, E. (1987a) 'The Financial Determinants of Technological Change. An Expository Survey', *Economic Notes*, vol. 16, no. 3, pp. 37–58.

Santarelli, E. (1987b) 'Generation and Diffusion of New Technologies', *International Review of Economics and Business*, vol. 34, no. 9, pp. 829–52.

Santarelli, E. (1991) 'Asset Specificity, R&D Financing, and the Signalling Properties of the Firm's Financial Structure', *Economics of Innovation and New Technology*, vol. 1, no. 4, pp. 279–94.

Santarelli, E. (1995) 'Directed Graph Theory and the Economic Analysis of Innovation', *Metroeconomica*, vol. 45, no. 2, pp. 111–126.

Santarelli, E. and E. Pesciarelli (1990) 'The Emergence of a Vision: the Development of Schumpeter's Theory of Entrepreneurship', *History of Political Economy*, vol. 22, no. 4, pp. 677–96.

Santarelli, E. and R. Piergiovanni (1994) 'Analysing Literature-based Innovation Output Indicators: The Italian Experience', Università di Bologna, Dipartimento di Scienze Economiche, Discussion Paper no. 197, *Research Policy*, forthcoming.

Santarelli, E. and A. Sterlacchini (1990) 'Innovation, Formal Vs. Informal R&D, and Firm Size: Some Evidence from Italian Manufacturing Firms', *Small Business Economics*, vol. 2, no. 3, pp. 223–8.

Santarelli, E. and A. Sterlacchini (1994a) 'New Firm Formation in Italian Industry: 1985–89', *Small Business Economics*, vol. 6, no. 2, pp. 95–106.

Santarelli, E. and A. Sterlacchini (1994b) 'Embodied Technological Change in Supplier Dominated Firms. The Case of Italian Traditional Industries', *Empirica*, vol. 20, no. 3, pp. 313–327.

Scherer, F. R. (1984) 'Invention and Innovation in the Watt–Boulton Steam Engine Venture', in *Innovation and Growth: Schumpeterian Perspectives* (Boston: MIT Press).

Schmookler, J. (1966) *Invention and Economic Growth* (Cambridge, Mass.: Harvard University Press).

Schmookler, J. (1972) *Patents, Invention and Economic Change* (Cambridge, Mass.: Harvard University Press).

Schumpeter, J. A. (1908) *Das Wesen und Der Hauptinhalt der Theoretischen Nationaloekonomie* (Leipzig: Duncker and Humblot).

Schumpeter, J. A. (1911) *Theorie der Wirtschaftlichen Entwicklung* (Leipzig: Duncker and Humblot).

Schumpeter, J. A. (1926) *Theorie der Wirtschaftlichen Entwicklung* (Leipzig: Duncker and Humblot).

Schumpeter, J. A. (1928) 'The Instability of Capitalism', *Economic Journal*, vol. 38, pp. 361–86.

Schumpeter, J. A. (1934) *Theory of Economic Development* (Cambridge, Mass.: Harvard University Press).

Schumpeter, J. A. (1939) *Business Cycles: A Theoretical, Historical, and Statistical Analysis of the Capitalist Process* (New York: McGraw-Hill), 2 vols.

Schumpeter, J. A. (1942) *Capitalism, Socialism, and Democracy* (New York: Harper).

Schumpeter, J. A. (1954) *History of Economic Analysis* (London: Allen & Unwin).

Servan-Schreiber, J. J. (1965) *The American Challenge* (Harmondsworth: Penguin).

Siegmann, K. (1990) 'Notebook Computers Challenge Desktops', *San Francisco Chronicle*, 19 December.

Silber, W. (ed.) (1975) *Financial Innovation* (Lexington: D. C. Heath).

Silber, W. (1983) 'The Process of Financial Innovation', *American Economic Review. Papers and Proceedings*, no. 65, pp. 89–95.

Simon, H. A. (1955) 'A Behavioural Model of Rational Choice', *Quarterly Journal of Economics*, vol. 69, no. 1, pp. 99–118.

Siniscalco, D. (1990) 'Evidenza empirica e aggregazione nelle teorie evolutive del cambiamento economico', in Amendola (ed.), pp. 113–18.

Smilor, R. W., D. V. Gibson and G. B. Dietrich (1990) 'University Spin-Out Companies: Technology Start-Ups from UT-Austin', *Journal of Business Venturing*, vol. 5, no. 1, pp. 63–76.

Solomou, S. (1987) *Phases of Economic Growth, 1850–1973. Kondratieff Waves and Kuznets Swings* (Cambridge: Cambridge University Press).

Soussou, H. (1985) 'Note on the Venture Capital Industry (1981)', mimeo, Harvard Business School.

Spence, A. M. (1973) *Market Signalling: Information Transfer in Hiring and Related Processes* (Cambridge, Mass.: Harvard University Press).

Spohr, A. P. and L. Wat (1985) 'What Investors Look for in R&D Partnerships', in Levine (ed.), pp. 41–4.

Sraffa, P. (1922) 'The Bank Crisis in Italy', *Economic Journal*, vol. 32, no. 126, pp. 178–97.

Stafford, D. C. and R. H. A. Purkis (eds) (1989) *Macmillan Directory of Multinationals* (London: Macmillan).

Stiglitz, J. (1972) 'Some Aspects of the Pure Theory of Corporate Finance: Bankruptcy and Takeovers', *Bell Journal of Economics*, vol. 3, no. 3, pp. 458–82.

Stiglitz, J. (1974) 'Incentives and Risk Sharing in Sharecropping', *Review of Economic Studies*, vol. 41, no. 2, pp. 219–55.

Stiglitz, J. (1988) 'Why Financial Structure Matters', *Journal of Economic Perspectives*, vol. 2, no. 2, pp. 121–26.

Stiglitz, J. and A. Weiss (1981) 'Credit Rationing in Markets with Imperfect Competition', *American Economic Review*, vol. 71, no. 3, pp. 393–410.

Stoneman, P. (1987) *The Economic Analysis of Technology Policy* (Oxford: Clarendon Press).

Stulz, R. M. (1990) 'Managerial Discretion and Optimal Financing Policies', *Journal of Financial Economics*, vol. 26, no. 1, pp. 3–27.

Suzuki, Y. (1987) *The Japanese Financial System* (Oxford: Clarendon Press).

Switzer, L. N. (1985) *The Financing of Technological Change* (Ann Arbor, Michigan: UMI Research Press).

Sylla, R. (1982) 'Monetary Innovation and Crises in American Economic History', in Wachtel (ed.).

Teece, D. (1986) 'Profiting From Technological Innovation: Implications for Integration, Collaboration, Licensing and Public Policy', *Research Policy*, vol. 15, no. 2, pp. 285–305.

Teece, D. J. (1989) 'Innovation and the Organization of Industry', mimeo, Center for Research in Management, University of California at Berkeley.

Tilly, R. (1986) 'German Banking 1850–1914: Development Assistance to the Strong', *Journal of European Economic History*, vol. 15, no. 1, pp. 113–52.

Tinbergen, J. (1951) *Econometrics* (London: George Allen & Unwin).

Tylecote, A. (1989) 'The South in the Long Wave: Technological Dependence and the Dynamics of World Economic Growth', in Di Matteo *et al.* (eds), pp. 206–24.

Tylecote, A. (1993) *The Long Wave in the World Economy. The Present Crisis in Historical Perspective* (London and New York: Routledge).

Tylecote, A. (1994) 'Long Waves, Long Cycles, and Long Swings', *Journal of Economic Issues*, vol. 28, no. 2, pp. 477–88.

Van Duijn, J. J. (1983) *The Long Wave in Economic Life* (London: George Allen & Unwin), first Dutch edition (1979) *De lange golf in de economie* (Assen: Van Gorcum).

Van Gelderen, J. (alias J. Fedder) (1913) 'Springvloed. Beschouwingen over industriele ontwikkeling en prijsbeweging', *De Nieuwe Tijd*, 18.

Von Ciriacy-Wantrup, S. (1936) *Agakrisen und Stockungsspannen. Zur Frage der langen Wellen in der wirtschaftlichen Entwicklung* (Berlin: Paul Parey).

Von Hippel, E. (1988) *The Sources of Innovation* (Oxford: Oxford University Press).

Von Hippel, E. (1989) 'Cooperation Between Rivals: Informal Know-How Trading', in Carlsson (ed.)

Von Tunzelmann, G. N. (1978) *Steam Power and British Industrialization to 1860* (Oxford: Oxford University Press).

Wachtel, P. (ed.) (1982) *Crises in the Economic and Financial Structure* (Lexington: D. C. Heath).

Walsh, V. (1984) 'Invention and Innovation in the Chemical Industry: Demand-pull or Discovery-push?', *Research Policy*, vol. 13, no. 4, pp. 211–34.

Williamson, O. E. (1975) *Markets and Hierarchies: Analysis and Antitrust Implications* (New York: The Free Press).

Williamson, O. E. (1981) 'The Modern Corporation: Origins, Evolution, Attributes', *Journal of Economic Literature*, vol. 19, no. 4, pp. 1537–68.

Williamson, O. E. (1985) *The Economic Institutions of Capitalism: Firms, Markets, Relational Contracting* (New York: The Free Press).

Williamson, O. E. (1988a) 'The Logic of Economic Organization', *Journal of Law, Economics, and Organization*, vol. 4, no. 1, pp. 65–93.

Williamson, O. E. (1988b) 'Corporate Finance and Corporate Governance', *Journal of Finance*, vol. 43, no. 3, pp. 567–91.

Williamson, O. E. (1991) 'Strategizing, Economizing, and Economic Organization', paper presented at the Conference on The New Science of Organization, Center for Research in Management, University of California at Berkeley, 18–20 January.

Index